ASCE Manuals and Reports on Engineering Practice No. 112

Pipe Bursting Projects

Prepared by
The Pipe Bursting Task Force of
the Trenchless Installation of Pipelines (TIPS)
Committee of the American Society of Civil Engineers

Edited by
Dr. Mohammad Najafi, P.E.

Library of Congress Cataloging-in-Publication Data

Pipe bursting projects / prepared by the Pipe Bursting Task Force of the Trenchless Installation of Pipelines (TIPS) Committee of the American Society of Civil Engineers ; edited by Mohammad Najafi.
p. cm. — (ASCE manuals and reports on engineering practice ; no. 112)
Includes bibliographical references and index.
ISBN 13: 978-0-7844-0882-7
ISBN 10: 0-7844-0882-3
1. Pipe bursting (Underground construction) 2. Underground pipelines—Maintenance and repair. 3. Trenchless construction. I. Najafi, Mohammad. II. American Society of Civil Engineers. Technical Committee on Trenchless Installation of Pipelines. Pipe Bursting Task Force.

TJ930.P4686 2006
621.8'672—dc22

2006027481

Published by American Society of Civil Engineers
1801 Alexander Bell Drive
Reston, Virginia 20191

www.pubs.asce.org

ISBN 13: 978-0-7844-0882-7
ISBN 10: 0-7844-0882-3
Manufactured in the United States of America.

MANUALS AND REPORTS ON ENGINEERING PRACTICE

(As developed by the ASCE Technical Procedures Committee, July 1930, and revised March 1935, February 1962, and April 1982)

A manual or report in this series consists of an orderly presentation of facts on a particular subject, supplemented by an analysis of limitations and applications of these facts. It contains information useful to the average engineer in his or her everyday work, rather than findings that may be useful only occasionally or rarely. It is not in any sense a "standard," however; nor is it so elementary or so conclusive as to provide a "rule of thumb" for nonengineers.

Furthermore, material in this series, in distinction from a paper (which expresses only one person's observations or opinions), is the work of a committee or group selected to assemble and express information on a specific topic. As often as practicable, the committee is under the direction of one or more of the Technical Divisions and Councils, and the product evolved has been subjected to review by the Executive Committee of the Division or Council. As a step in the process of this review, proposed manuscripts are often brought before the members of the Technical Divisions and Councils for comment, which may serve as the basis for improvement. When published, each work shows the names of the committees by which it was compiled and indicates clearly the several processes through which it has passed in review, in order that its merit may be definitely understood.

In February 1962 (and revised in April 1982) the Board of Direction voted to establish a series entitled "Manuals and Reports on Engineering Practice," to include the Manuals published and authorized to date, future Manuals of Professional Practice, and Reports on Engineering Practice. All such Manual or Report material of the Society would have been refereed in a manner approved by the Board Committee on Publications and would be bound, with applicable discussion, in books similar to past Manuals. Numbering would be consecutive and would be a continuation of present Manual numbers. In some cases of reports of joint committees, bypassing of Journal publications may be authorized.

MANUALS AND REPORTS ON ENGINEERING PRACTICE

No.	Title
13	Filtering Materials for Sewage Treatment Plants
14	Accommodation of Utility Plant Within the Rights-of-Way of Urban Streets and Highways
35	A List of Translations of Foreign Literature on Hydraulics
40	Ground Water Management
41	Plastic Design in Steel: A Guide and Commentary
45	Consulting Engineering: A Guide for the Engagement of Engineering Services
46	Pipeline Route Selection for Rural and Cross-Country Pipelines
47	Selected Abstracts on Structural Applications of Plastics
49	Urban Planning Guide
50	Planning and Design Guidelines for Small Craft Harbors
51	Survey of Current Structural Research
52	Guide for the Design of Steel Transmission Towers
53	Criteria for Maintenance of Multilane Highways
54	Sedimentation Engineering
55	Guide to Employment Conditions for Civil Engineers
57	Management, Operation and Maintenance of Irrigation and Drainage Systems
59	Computer Pricing Practices
60	Gravity Sanitary Sewer Design and Construction
62	Existing Sewer Evaluation and Rehabilitation
63	Structural Plastics Design Manual
64	Manual on Engineering Surveying
65	Construction Cost Control
66	Structural Plastics Selection Manual
67	Wind Tunnel Studies of Buildings and Structures
68	Aeration: A Wastewater Treatment Process
69	Sulfide in Wastewater Collection and Treatment Systems
70	Evapotranspiration and Irrigation Water Requirements
71	Agricultural Salinity Assessment and Management
72	Design of Steel Transmission Pole Structures
73	Quality in the Constructed Project: A Guide for Owners, Designers, and Constructors
74	Guidelines for Electrical Transmission Line Structural Loading
76	Design of Municipal Wastewater Treatment Plants
77	Design and Construction of Urban Stormwater Management Systems
78	Structural Fire Protection
79	Steel Penstocks
80	Ship Channel Design
81	Guidelines for Cloud Seeding to Augment Precipitation
82	Odor Control in Wastewater Treatment Plants
83	Environmental Site Investigation
84	Mechanical Connections in Wood Structures
85	Quality of Ground Water
86	Operation and Maintenance of Ground Water Facilities
87	Urban Runoff Quality Manual
88	Management of Water Treatment Plant Residuals
89	Pipeline Crossings
90	Guide to Structural Optimization
91	Design of Guyed Electrical Transmission Structures
92	Manhole Inspection and Rehabilitation
93	Crane Safety on Construction Sites
94	Inland Navigation: Locks, Dams, and Channels
95	Urban Subsurface Drainage
96	Guide to Improved Earthquake Performance of Electric Power Systems
97	Hydraulic Modeling: Concepts and Practice
98	Conveyance of Residuals from Water and Wastewater Treatment
99	Environmental Site Characterization and Remediation Design Guidance
100	Groundwater Contamination by Organic Pollutants: Analysis and Remediation
101	Underwater Investigations
102	Design Guide for FRP Composite Connections
103	Guide to Hiring and Retaining Great Civil Engineers
104	Recommended Practice for Fiber-Reinforced Polymer Products for Overhead Utility Line Structures
105	Animal Waste Containment in Lagoons
106	Horizontal Auger Boring Projects
107	Ship Channel Design (Second Edition)
108	Pipeline Design for Installation by Horizontal Directional Drilling
109	Biological Nutrient Removal (BNR) Operation in Wastewater Treatment Plants
110	Sedimentation Engineering: Processes, Measurements, Modeling, and Practice
111	Reliability-Based Design of Utility Pole Structures
112	Pipe Bursting Projects

CONTENTS

PREFACE

This Manual of Practice (MOP) was prepared by the Pipe Bursting Task Force of the ASCE Committee on Trenchless Installation of Pipelines (TIPS), under supervision of the Pipeline Division. This manual describes current pipe bursting practices used by engineers and construction professionals in designing and constructing pipelines under roads, railroads, streets, and other man-made and natural structures and obstacles. The Trenchless Installation of Pipelines (TIPS) Committee under leadership of Dr. Ahmad Habibian, P.E. (Past Chair) and Mr. Timothy Stinson, P.E. (Current Chair) is credited for the efforts leading to this publication. The committee would like to thank contributors, task committee members, and blue ribbon reviewers, whose names follow, for their support, time, and efforts.

Mohammad Najafi
ASCE Pipe Bursting Task Committee Chair

ACKNOWLEDGMENTS

Contributors
Part 1: General
Team Leader: Alan Atalah, Bowling Green State University
Dennis Doherty, Jacobs Associates
Tim Stinson, S E A Consultants, Inc.
Tony Almeida, Halff Associates
Henry Nodarse, Bernardin, Lochmueller & Associates, Inc.

Part 2: Planning Phase
Team Leader: Terry McArthur, HDR Engineering, Inc.
Robert Carpenter, Underground Construction (Oildom Publishing)
Brett Affholder, Insituform Technologies
Randy Robertson, Cyntech Corp.
Dave Kozman, RS Lining Systems, LLC
Peter Funkhouser, Michigan Department of Transportation

Part 3: Existing (Host) Pipe
Team Leader: Ralph Carpenter, American Ductile Iron Pipe/American Spiral Weld Pipe
Larry Petroff, Performance Pipe
Tom Marti, Underground Solutions, Inc.
Shah Rahman, S&B Technical Products/Hultec
Larry Slavin, Outside Plant Consulting Services, Inc.

Part 4: New (Replacement) Pipe
Team Leader: Larry Slavin, Outside Plant Consulting Services, Inc.
Ralph Carpenter, American Ductile Iron Pipe/American Spiral Weld Pipe
Larry Petroff, Performance Pipe
Tom Marti, Underground Solutions, Inc.
Shah Rahman, S&B Technical Products/Hultec

Part 5: Design and Preconstruction Phase
Team Leader: Terry Moy, Woolpert, Inc.
Mark Dionise, Michigan Department of Transportation

George Davis, Missouri Department of Transportation
Brian Dorwart, Haley & Aldrich, Inc.

Part 6: Construction Phase
Team Leaders: Ben Cocogliato, TT Technologies, Inc. and
Eric Nicholson, Hammer Head Mole
Mike Mason, Nowak Construction Company
Al Tenbusch, Tenbusch, Inc.
Oleh Kinash, Center for Underground Infrastructure Research and Education (CUIRE)
Abdel Tayebi, Center for Underground Infrastructure Research and Education (CUIRE)
Brian Hunter, TT Technologies, Inc.
Alan Goodman, Hammer Head Mole
Jeff Wage, Hammer Head Mole
Dave Holcomb, TT Technologies, Inc.

Blue Ribbon Reviewers
David Bennett, Bennett/Staheli Engineers
Tennyson M. Muindi, Haley & Aldrich, Inc.
Collins Orton, TT Technologies, Inc.

ASCE Representatives
John Segna, Director, Technical Activities
Verna Jameson, Senior Coordinator, Technical Activities
Suzanne Coladonato, Manager, Book Production

Task Committee Officers
Chair: Mohammad Najafi, Center for Underground Infrastructure Research and Education (CUIRE)

Vice Chair: Ralph Carpenter, American Ductile Iron Pipe/American Spiral Weld Pipe

Secretary: Craig Camp, Jacobs Associates

Pipeline Division Executive Committee
Dr. Tom Iseley, P.E., Chair
Joe Castronovo, P.E.
Dr. Ahmad Habibian, P.E.
William J. Moncrief, P.E.
Dr. Mohammad Najafi, P.E.
Michael T. Stift, P.E.
Randy Robertson, P.E.

PART 1

GENERAL

1.1 INTRODUCTION

Pipe bursting is a well-established trenchless method that is widely used for the replacement of an existing and deteriorated pipe with a new pipe of the same or larger diameter. Many factors should be reviewed thoroughly before pipe bursting projects are considered and released for bid. Engineers should consider different options and select the most cost-effective and environmentally friendly methods for bid. The method selection should not be left to only the judgment of the contractor. This manual will help engineers and owners in the method selection process.

Pipe bursting is especially cost-effective if the existing pipe is out of capacity. This method can be used advantageously to reduce damage to pavements and disruptions to traffic, hence reducing the social costs associated with pipeline installations. There are, however, limits to the use of the pipe bursting method, and various conditions challenge the successful use of its application. This manual provides information that is essential for the engineer as well as the contractor for the successful and safe execution of pipe bursting projects.

Although the pipe bursting method is commonly used for replacing an existing pipe, it has not been covered adequately by manuals, guidelines, or standards. The need to develop a comprehensive manual for pipe bursting projects arose as a result of steady advancements in the field and the lack of proper engineering guidelines. This manual, developed by the Pipe Bursting Task Force of the ASCE Committee on Trenchless Installation of Pipelines (TIPS), is a major step toward promoting best practices and creating a knowledge base for pipe bursting projects. This manual will assist engineers, contractors, and owners in designing and carrying out projects effectively and safely in conformance with project requirements and site conditions by providing an understanding of the method and its capabilities and limitations. In addition, this manual outlines design and construction considerations and identifies potential problems and prevention measures, thereby instilling confidence in the appropriate use of the method. These guidelines are based on information obtained from manufacturers' literature, technical papers and other related information, and from comments and reviews made by industry experts.

1.2 HISTORY OF PIPE BURSTING DEVELOPMENT

Pipe bursting was first developed in the UK in the late 1970s by D. J. Ryan & Sons in conjunction with British Gas, for the replacement of small-diameter, 3- and 4-in. (75- and 100-mm) cast iron gas mains (Howell 1995). The process involved a pneumatically driven, cone-shaped bursting head operating by a reciprocating impact process. This method was patented in the UK in 1981 and in the United States in 1986. While the original patents expired in April, 2005, new proprietary pipe bursting methods have been developed and patented. When it was first introduced, this method was used only in replacing cast iron gas distribution lines; it was later employed to replace water and sewer lines. By 1985, the process was further developed to install up to 16-in. (400-mm) outer diameter (OD) medium-density polyethylene (MDPE) sewer pipe. Replacement of sewers in the UK using sectional pipes as opposed to continuously welded polyethylene pipe was described in a paper by Boot et al. (1987). The total footage of pipe replaced using pipe bursting in the United States is growing at approximately 20% per year, the majority of which is for sewer line replacement.

1.3 WHAT IS PIPE BURSTING?

Pipe bursting is defined as a replacement method in which an existing pipe is broken by brittle fracture, using mechanically applied force from within. The pipe fragments are forced into the surrounding ground. At the same time, a new pipe of the same or larger diameter is drawn in, replacing the existing pipe. Pipe bursting involves the insertion of a conically shaped tool (bursting head) into the existing pipe to shatter the existing pipe and force its fragments into the surrounding soil by pneumatic or hydraulic action. A new pipe is pulled or pushed in (depending on the type of the new pipe) behind the bursting head.

The base of the bursting head is larger than the inside diameter (ID) of the existing pipe and slightly larger than the outside diameter (OD) of the new pipe to reduce friction and to provide space for maneuvering the pipe. The back end of the bursting head is connected to the new pipe and the front end is connected to a cable or pulling rod. The new pipe and bursting head are launched from the insertion shaft and the cable or pulling rod is pulled from the pulling shaft, as shown in Fig. 1-1. The bursting head is usually attached to other components to lengthen the bursting body to reduce the effects of sags or misalignment on the new pipeline. Some bursting tools are equipped with expanding crushing arms, sectional ribs, or sharp blades to transfer point or line loads to the existing pipe to assist in bursting. To break the existing pipe, the bursting head receives pulling force from the pulling cable or rods and hydraulic or pneumatic power to the head, based on the bursting system being used. The pulling force is transferred to the existing pipe, breaking it into pieces and expanding the diameter of the cavity. The bursting head is pulled through the pipe debris, creating a cavity and pulling behind it the new pipe from the insertion shaft. The basic differences between these systems are in the source of pulling force and some consequent differences in operation.

The pipe bursting method creates a cavity in the soil around the pipe where the new pipe is pulled through. The cavity creates a compression plastic zone around the new pipe, outlined by an elastic zone as shown in Fig. 1-2. The magnitude of the compression and the dimensions of these zones correlate with the amount of upsizing, the diameter of the pipe, and the type of soil.

Underground service utilities in many cities have been in place for more than 100 years. Although existing systems have functioned well beyond reasonably anticipated service life, underground systems are now mostly deteriorated and need costly maintenance and repair. Common problems involve corrosion and deterioration of pipe materials, failure or leakage of pipe joints, and reduction of flow due to mineral deposits and debris buildup inside the pipe. Damage can also occur by ground movements due to adjacent construction activity, earthquakes, or relative movements caused by differential settlement or other ground instability.

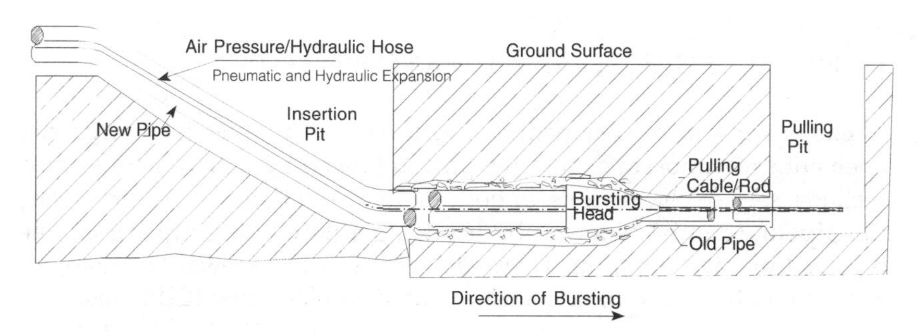

FIGURE 1-1. Pipe bursting operation.

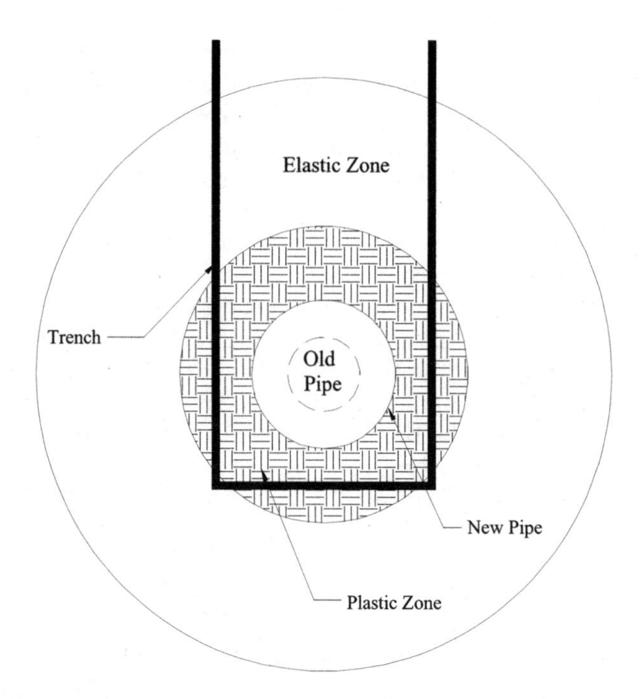

FIGURE 1-2. Cavity expansion and the plastic and elastic zones.

This leads to an increase in infiltration and inflow (I&I) in sewer systems and to reduced flows and pressures, persistent leakage (of up to 30% of water provided in some systems), pipe bursts, and poor water quality in water systems. These problems tend to increase with the age of the network. Maintaining this large network of underground sewer, water, and gas pipelines is difficult and costly. The problems are compounded by the significant negative impacts that open cut repairs or replacement projects have on the daily life, traffic, and commerce of the area served by and along the pipeline in question.

1.4 PIPE BURSTING FEASIBILITY AND BENEFITS

For repair and replacement, conventional techniques have involved open cut excavation to expose the pipe, followed by replacement of pipe sections, service connections, or both for localized damage. Alternatively, the pipeline can be renewed by inserting a new lining pipe or can be replaced by pipe bursting. There are several pipe renewal technologies available in the marketplace, such as cured-in-place pipe (CIPP), close-fit pipe, and modified sliplining. The main advantage of some of these renewal methods over pipe bursting is that they may not need excavation for access or lateral connections.

In contrast, pipe bursting has the advantage of maintaining or even increasing the gravity pipeline capacity by more than 100%. After some algebraic manipulation to the following Chezy-Manning equation (which is used to calculate the capacity of gravity lines), it is easy to prove that a 15% or 32% increase in the pipe ID, combined with the smoother pipe surface, can produce a 100% or 200% increase in the flow capacity, respectively.

$$Q = \frac{1.49}{n} A(r_H)^{2/3} \sqrt{S} \tag{1-1}$$

where

Q = the volume flow rate (ft^3/sec),
n = Manning roughness coefficient (unitless),
A = the cross-sectional area of the pipe (ft^2),
r_H = hydraulic radius (ft), and
S = the slope of the energy line (unitless), which is parallel to the water surface and pipe invert if the flow is uniform.

The n value for new clay or concrete pipes ranges between 0.012 and 0.015 (on average, about 0.013), and is about 0.009 for high-density polyethylene (HDPE) and polyvinyl chloride (PVC) pipes. Therefore, a smoother pipe with the same ID can produce a 44% increase in flow (Lindeburg 1992). A similar analysis can be presented for pressure flow.

Although each project is unique and cost analysis for different project and site conditions and applicable methods should be performed on case-by-case basis, it has been shown that pipe bursting *generally* has substantial advantages over open cut replacements. Pipe bursting is usually much faster, more efficient, and often less expensive than open cut (especially in sewer line replacement) because of the depth of the sewer lines. The increased sewer depth may require extra excavation, shoring, and dewatering, which substantially increases the cost of open cut replacement. The increased depth has a minimal effect on the cost per foot for pipe bursting, as shown in Fig. 1-3. Specific studies in the United States have shown that pipe bursting cost savings as high as 44% (with average savings of 25%) compared to open cut were achievable (Fraser et al. 1992). Cost savings could be much higher in difficult ground conditions because excavation in such conditions becomes extremely expensive compared to pipe bursting. In addition, open cut can cause significant damage to nearby buildings and structures.

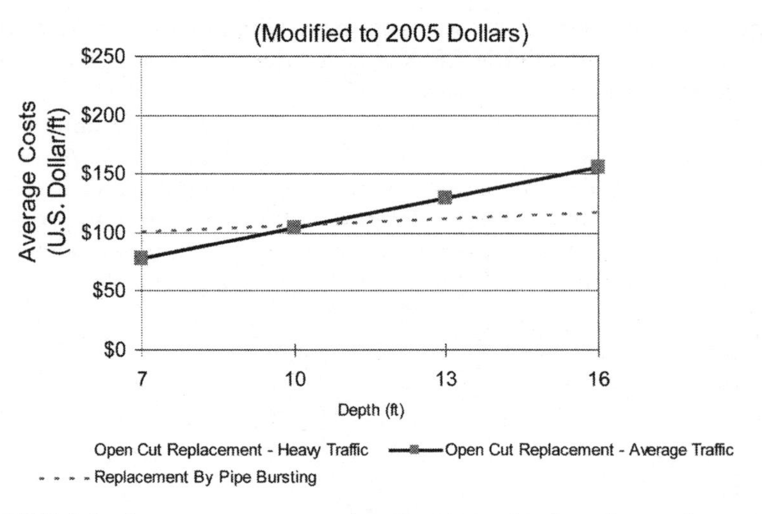

FIGURE 1-3. Average cost comparison between pipe bursting and open cut replacements [modified from Poole et al. (1985)].

In addition to the direct cost advantage of pipe bursting over open cut, the trenchless technique of pipe bursting has several indirect (social) cost savings. Less time for replacement and fewer traffic disturbances, road or lane closures, business interruptions, and environmental intrusions are some examples of these social cost savings.

As mentioned previously, the unique advantage of pipe bursting over pipe lining techniques such as CIPP, sliplining, and deform and reform is the ability to upsize the service lines. A 30% upsizing doubles the capacity of the sewer line, excluding the effect of the smoother pipe. The technique is most cost-advantageous compared to the lining techniques when (1) there are few lateral connections to be reconnected within a replacement section; (2) the existing pipe is structurally deteriorated; and (3) additional capacity is needed.

On the other hand, currently pipe bursting has the following limitations: (1) excavation for the lateral connections is needed; (2) expansive soils could cause difficulties for bursting; (3) a collapsed pipe at a certain point along the pipe run may require excavation at that point to allow the insertion of pulling cable or rod and to fix the pipe sag; (4) encasement, sleeves, and point repairs with ductile material can interfere with the replacement process; (5) if the existing sewer line is significantly out of line and grade, the new line will also tend to be out of line and grade, although some corrections of localized sags are possible; and (6) insertion and pulling shafts are needed. Existing pipe fittings may or may not create problems, depending on the type and size of the existing pipe.

1.5 APPLICABILITY

Pipe bursting can be applied on a wide range of pipe sizes and types and in a variety of soil and site conditions. The size of pipes being burst typically ranges from 2 to 30 in. (50 to 750 mm), although pipes of larger sizes can be burst. Pipe bursting is commonly done size-for-size and one size upsize above the diameter of the existing pipe. Larger upsizes (up to three pipe sizes) have been successful, but the larger the pipe upsizing, the more pulling force is needed and the more ground movement may be experienced.

Different site and project conditions are experienced during every bursting activity. It is important to pay close attention to the project surroundings, depth of bury, and soil conditions when replacing an existing pipe using pipe bursting. It is also important to determine a minimum depth of cover (based on existing pipe diameter, the original trench conditions, and possible upsizing) to prevent surface and pavement heaving. Some unfavorable conditions are expansive soils, repairs made with ductile material, collapsed pipe, concrete encasement, sleeves, and adjacent pipes or utility lines. These conditions require extra attention to ensure the safety of all involved people as well as the surrounding facilities and infrastructure.

In all applications of pipe bursting it is necessary to know the soil conditions, the location of all surrounding structures and infrastructure, and the effects of pipe bursting on these conditions. Pipe bursting operations can be done safely if its compatibility with site and project conditions is determined with certainty.

1.6 PIPE BURSTING SYSTEMS

To break the existing pipe, the bursting head receives pulling force from the pulling cable or rods, hydraulic power to the head, or pneumatic power to the head, according to the bursting system. Existing mains can be replaced by one of several trenchless techniques developed to date: pneumatic, static pull, hydraulic expansion, implosion, pipe splitting, pipe pushing, or pipe grinding. The selection depends on ground conditions, groundwater conditions, degree of upsizing required, type of new pipe, construction of the existing pipeline, and depth of the pipeline. The basic differences between these systems are in the source of pulling force, the method of breaking the existing pipe, and some consequent differences in operation that are briefly described in the following paragraphs.

1.6.1 Pneumatic Bursting Systems

In the pneumatic bursting system, the bursting tool is a soil displacement hammer driven by compressed air and operated at a rate of 180 to

(A) (B)

FIGURE 1-4. The bursting head (A) and winch (B) for a pneumatic pipe bursting system.

580 blows per minute. FIGURE 1-4 presents the bursting head (A) and winch (B) for this system. Note that the winch is positioned at the pulling shaft. The percussive action of the hammering cone-shaped head is similar to hammering a nail into the wall (each hammer blow pushes the nail a short distance). With each stroke it cracks and breaks the existing pipe. The expander on the head, combined with the percussive action, pushes the fragments and the surrounding soil away, providing space for the new pipe. The tension applied to the cable keeps the bursting head aligned with the existing pipe, keeps the bursting tool pressed against the existing pipe wall, and pulls the new pipe behind the head. An air pressure supply hose is inserted through the new pipe and is connected to the bursting tool. Once the head is attached to the new pipe, the winch cable is inserted through the existing pipe and attached to the head; the air compressor and the winch are set at a constant pressure and tension values, respectively, and the bursting head process starts. The process continues with little operator intervention until the head reaches the pulling shaft, at which point it is separated from the new pipe.

For size-on-size and single upsize bursts up to 12-in. diameter, front-mount pneumatic bursting is the most common method used with fracturable pipes. This allows for manhole exiting, thus eliminating the need for a second pit. Upon burst completion, the burst head is removed from inside the manhole and exited through the manhole cover. Eliminating a second pit, when applicable, can produce significant cost savings for the project owner.

1.6.2 Static Bursting Systems

In the static pull system, no hammering action is used. A large pull force is applied to the cone-shaped expansion head by a pulling rod assembly or cable (winch) inserted through the existing pipe. The cone transfers the horizontal pulling force into a radial force, breaking the existing pipe and expanding the cavity to provide space for the new pipe. FIGURE 1-5 presents an example bursting head and winch for the static pull pipe bursting system. Steel rods, each about 4 ft (1.2 m) long, are inserted into the existing pipe from the pulling shaft. The rods are connected using different types of connections. When the rods reach the insertion shaft, the bursting head is connected to the rods and the new pipe is connected to the rear of the head. A hydraulic unit in the pulling shaft pulls the rods one rod at a time and the rod sections are removed. The bursting head and the new pipe are pulled with the rod or the winch, fracturing the existing pipe and pushing the debris to the surrounding soil. The process continues until the bursting head reaches the pulling shaft, where it is separated from the new pipe. If a cable or winch is used instead of a rod assembly, the pulling process continues with minimum interruption, but less force is available for the operation.

1.6.3 Hydraulic Bursting Systems

Topf (1991; 1992) and Tucker et al. (1987) described a hydraulic system for pipe bursting in which the bursting head is sequentially pulled through

FIGURE 1-5. The bursting head and winch for a static pull pipe bursting system.

the pipe to the desired location and then expanded laterally to break the pipe. Steel mains have been burst using a combination of splitter wheels, a sail blade, and an expander unit (Fisk and Zlokovitz 1992). A direct, continuous (either pull or push) hydraulic system is also commonly employed using a cone-shaped bursting head and a rod system for applying the tensile or thrust force.

In a hydraulic expansion system, the bursting head is constructed in the form of four or more interleaved segments, hinged at the ends and at the middle, as shown in Fig. 1-6. The head expands and contracts through the action of an axially mounted hydraulic piston. The head is pulled with a cable running through the existing pipe by a winch from the pulling shaft. It is also connected to the new pipe from the back end. The hydraulic supply lines are inserted through the new pipe and are connected to the head. Once the head is inserted through the existing pipe and pulled by the winch into position within the interior of the pipe wall, the head expands, breaking the existing pipe and compressing the fragments into the surrounding soil. The contraction of the head and the tension from the winch allow the head to move forward, pulling the new pipe behind it until the head is in position to break the next segment. The process is repeated until the head travels from the insertion shaft to the pulling shaft.

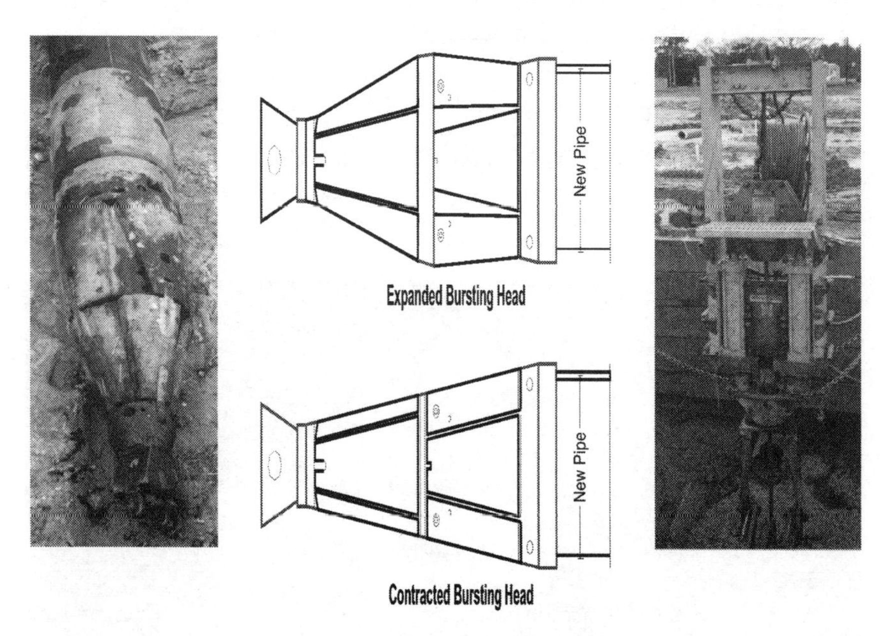

FIGURE 1-6. The bursting head and winch for a hydraulic pipe bursting system.

1.6.4 Tenbusch Insertion Method

The Tenbusch Insertion Method (TIM™) pushes or jacks new pipe into the existing deteriorated pipe. The TIM™ system utilizes the columnar strength of segmented bell-less jacking pipe to advance the lead train through the existing pipe. The lead train consists of five sections (Fig. 1-7):

1. The **lead**—a heavy steel guide pipe that maintains the alignment within the center of the existing pipe.
2. The **cracker**— fractures the existing pipe.
3. The **cone expander**—radially expands the fractured line into the surrounding soil.
4. The **front jack**—a hydraulic cylinder that provides axial thrust to the (above) penetration/compaction pieces.
5. The **pipe adapter**—provides mating surfaces which link the new pipe to the front jack.

The last section, the **pipe adapter**, is fitted with a lubricant injection port where lubricant (polymer or bentonite) can be injected into the annular space surrounding the new replacement pipe; the introduction of a lubricant allows for the efficient replacement of existing pipe lines even in soft, sticky clays or wet sands.

Dual flexible hose sections that transport lubricant and hydraulic fluid to the front train are fed through each new pipe section. Each new hose section is connected to previous sections and to the operator's control panel with quick-disconnect couplings. Using the new pipe as a support column, the front jack advances the lead train into the existing pipe independent of the advance of the new pipe column. The new pipe is jacked

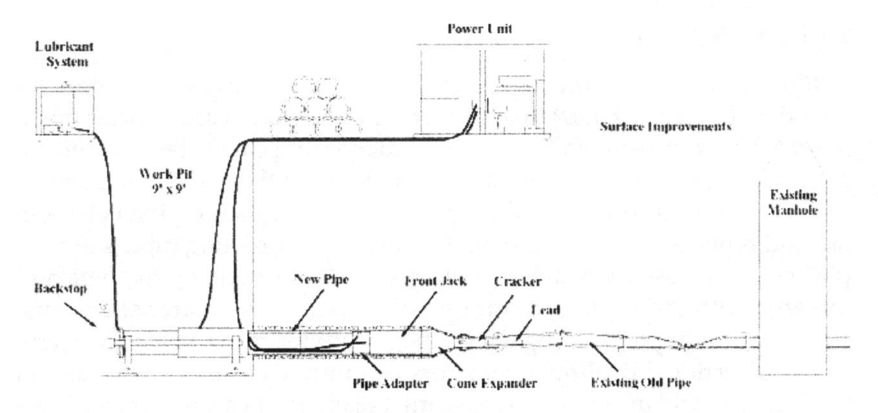

FIGURE 1-7. The Tenbusch Insertion Method™.

behind the lead train piece-by-piece by the jacking frame in the work pit. The primary jacking frame applies the required thrust to advance the new pipe column as the front jack is retracted. Instrumentation and controls at the operator's control panel (at the jacking frame) allow the operator to feel his way through the existing pipe as the new pipe column and front train are "inchwormed" into the existing line. Upon completion of the line replacement, the lead train is disassembled inside a typical 4-ft-diameter (~1.2-m) receiving manhole and the new pipe is jacked into its final position.

1.6.5 Pipe Splitting

Pipe splitting is a bursting method for breaking an existing pipe by longitudinal splitting. At the same time, a new pipe of the same or larger diameter may be drawn in behind the splitting tool. Pipe splitting is used to replace ductile pipe materials, which do not fracture using the above-cited bursting techniques. The system has a splitting wheel or cutting knives that slit the pipe longitudinally at two or more lines along the side of the pipe.

1.7 TRENCHLESS PIPELINE REMOVAL SYSTEMS

Although pipe bursting technology breaks the existing pipe and *pushes* the particles into surrounding earth, pipe removal systems (as their name implies) break the existing pipe but *remove* the particles out of the ground. There are several methods of pipe removal, as described in the following sections.

1.7.1 Pipe Reaming

Pipe reaming is a pipe replacement technique that uses a horizontal directional drilling (HDD) machine with minor modification. After pushing the drill rods through the existing pipeline and connecting the rods to a special reamer, the new pipe string is attached to the reamer via a swivel and towing head. As the drill rig pulls back, the existing pipe is broken out and replaced by the new pipe. Removal of the existing pipe is accomplished by the use of an appropriate reamer that grinds up the pipe and mixes it with the drilling fluid. The pulverized pipe material is transferred to an exit point (using the drilling fluid) for removal via a vacuum truck. Directional drilling contractors or utility contractors who use an HDD rig can add inexpensively modified reamers of various types for different pipe materials and ground conditions. According to Nowak Pipe Reaming, Inc., the patent holder for the InneReam System, this technique is limited to nonmetallic pipeline replacement where surrounding environmental conditions such as expansive soils, rock, and concrete encasement that may prohibit other procedures are not obstacles to successful installations. Dropped joints, protruding service taps or collapsed pipes, misalignment or sags, vertical sags, joint deflections, and protrusions are largely remedied by the use of an extended mandrel on the reamer. A collapsed section, if severe, may require an open cut to allow insertion of the drill rod.

Pipe reaming is not a suitable process for replacing existing cast iron (CI) or ductile iron pipes (DIP). However, using the proper reamer, Class III reinforced concrete pipe (RCP) has successfully been replaced by this method. Because the HDD method is used, the contractor should have a contingency plan to provide immediate relief if mud return is lost or evidence of inadvertent drilling fluid returns known as "frac'ing out" is discovered. Also for this method, service connections should be disconnected and capped *before* an installation begins to avoid filling the service line with drilling fluids.

FIGURE 1-8 illustrates pipe reaming, which is an adaptation of back reaming using an HDD rig. The HDD [figure label (1)] is located over and preferably in line with the pipeline to be replaced (4), at a distance from the point of entry into the new line commensurate with the safe minimum bending radius of the drill rod (2). The drill rod (2) is extended through the existing pipe (4) into the insertion pit where a specially designed reamer (3) and the replacement pipe (7) are attached. The existing pipe (4) is ground, pulverized, mixed with the drilling fluids, and flushed through the existing pipe (4) to a manhole, service line excavation, or excavated pit where it is retrieved by a vacuum truck for disposal. The replacement pipe is installed simultaneously as the reaming progresses.

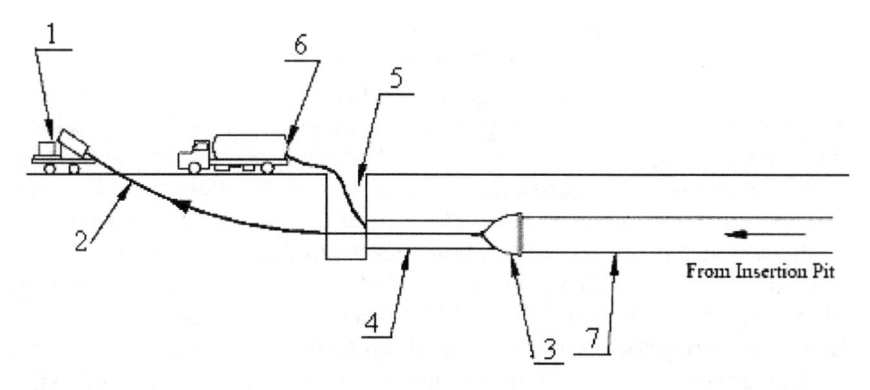

FIGURE 1-8. A pipe reaming system.

1.7.2 Air Impactor

Air impactor pipe bursting systems use hammer technology to merge the capabilities of pneumatic and static bursting systems. Great tensile force is used to overcome pipe friction and engage the hammer. The hammering action of the air impactor breaks the existing pipe and expands the cavity, providing space for the new pipe. This hammer operates independently from the pulling rod. The steel rods (usually HDD rods) are inserted into the existing pipe by drilling down to a manhole with an HDD rig. When the rods reach the insertion shaft, the air impactor is connected to the rods and the new pipe is connected to the rear of the air impactor via an HDPE adaptor. The air impactor is powered pneumatically through a front-feed air connection through the drill string, or is fed through a rear connection using an air hose. As the pulling unit pulls the rods back, the friction opens a valve which activates the air impactor. When the pulling stops, the air impactor goes into a vent mode and vents air through the back end of the tool. This process continues until the air impactor reaches the pulling shaft where it is separated from the new pipe. This system was originally designed for use with existing HDD technology. Using HDD technology in challenging high-traffic setup areas often enables a crew to work in an easement off to the side, further minimizing traffic disruption. As mentioned previously, using an HDD drill string for air supply can minimize labor on a project. Air impactor also known as smart hammer technology has been proven effective with a variety of pullback systems, including cable winches and static pipe bursting machines that use torqued joint rods. When these alternative methods of connecting to an air impactor are used, an air hose must be fed through the new product pipe to feed the air impactor from the rear. FIGURE 1-9 illustrates the air impactor technology.

1.7.3 Pipe Eating

Pipe eating is a rarely used, modified microtunneling system specially adapted for pipe replacement. The microtunnel boring machine (MTBM) crushes the existing defective pipeline and removes the particles by a circulating slurry system. A new pipe is simultaneously installed by jacking it behind the microtunneling machine. The new pipe may follow the line of the old pipe on the entire length, or it may cross the elevation of the old pipe on a limited segment only. The microtunneling system is remotely controlled and is guided with a laser line from the drive pit. During the operation, the MTBM removes or *eats* whatever is in its way—the existing pipe or the ground itself. The MTBM has a cuttinghead and a shield section. The cuttinghead has cutting teeth and rollers that cut the pipe, and cutting arrangements close to the edge of the shield cut the ground to the required diameter to take the new pipe. The cuttinghead is cone-shaped,

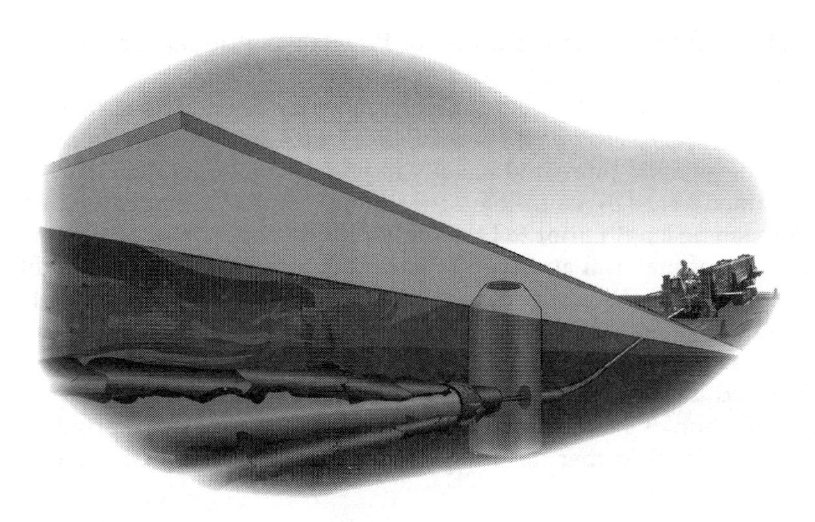

FIGURE 1-9. The air impactor technology.

which puts the existing pipe material into tension and thus reduces heavy wear on the cutting teeth. The shield section carries the cuttinghead and its hydraulic motor system. The MTBM is launched from a drive pit, where a thrust frame is located. The jacking frame provides a thrust force that is applied to the new pipe to push the MTBM forward.

1.8 PIPE MATERIAL

Existing pipe is typically made of a rigid material such as vitrified clay pipe (VCP), CI, plain concrete, asbestos cement, or certain plastics. RCP can be successfully replaced when it is not heavily reinforced or if it is substantially deteriorated. The size of the existing pipe typically ranges from 2 to 30 in. (50 to 750 mm), although the bursting of larger diameters is increasing. A length of 300 to 400 ft (90 to 120 m) is typical for bursting; however, much longer runs have been completed with more powerful pieces of equipment. It should be noted that point repairs on the existing pipe (especially repairs made with ductile materials) can make the process more difficult.

HDPE and MDPE pipes have been the most commonly used replacement pipes for pipe bursting applications. The main advantages of polyethylene (PE) pipe are its continuity, flexibility, and versatility. Pipe string is obtained by heat-fusing long segments together in the field; producing continuity during the installation reduces the need for stopping the process. Flexibility allows bending the pipe for angled insertion in the field. In addition, the versatility of PE enables it to meet all the other requirements for gas, water, and wastewater lines.

The new pipe can be made of many other pipe materials, such as PVC, DIP, VCP, RCP, FRP, or PCP.[1] VCP, RCP, FRP, and PCP are collectively called segmental pipes and usually require jacking operation. However, restrained joint DIP or fusible or restrained joint PVC can be assembled into a single length prior to bursting for pulling operation, depending on the type of material and type of joint they use. When pipes cannot be pulled in, these pipes can be jacked into position behind the bursting head or kept compressed by towing them via a cap connected to the cable or rod that passes through the pipes. The joints of these pipes must be designed for trenchless installations. Therefore, the static pull system is the only bursting system that can be used with these pipes. Parts 3 and 4 of this manual further discuss pipe material for both existing and new (replacement) pipes.

1.9 SCOPE OF THIS MANUAL

This manual presents facts and guidelines on pipe bursting technology, supplemented by an analysis of limitations and applications of these facts. Part 2 of this manual presents the planning phase, Part 3 presents existing pipe materials, Part 4 presents new replacement pipe materials, Part 5 presents the design and preconstruction aspects, Part 6 of this manual presents the construction phase, and Part 7 presents the list of references.

1.10 RELATED DOCUMENTS

American Society of Civil Engineers (ASCE). (1996). "Pipeline crossings." *ASCE Manual and Reports on Engineering Practice No. 89*, Reston, Va.

American Society of Civil Engineers (ASCE). (2004). "Horizontal auger boring projects." *ASCE Manual and Reports on Engineering Practice No. 106*, Reston, Va.

American Society of Civil Engineers (ASCE). (2005). "Pipeline Design for Installation by Horizontal Directional Drilling." *ASCE Manual and Reports on Engineering Practice No.* 108, Reston, Va.

ASCE (1997). "Geotechnical baseline reports for underground construction," American Society of Civil Engineers, Reston, VA, *ASCE*, 0-7844-0249-3, 1997.

---

[1]Pipe abbreviations: PVC: Polyvinyl chloride; DIP: Ductile iron pipe; VCP: Vitrified clay pipe, RCP: Reinforced concrete pipe; FRP: Fiberglass reinforced polyester mortar; and PCP: Polymer concrete pipe.

ASCE. (2002). "Standard guidelines for the collection and depiction of existing subsurface utility data." *CI/ASCE Standard 38-02*, Reston, Va.

Atalah, A. (1998). "The effect of pipe bursting on nearby utilities, pavement, and structures." *Technical Report TTC-98-01*, Trenchless Technology Center, Louisiana Tech University, Ruston, La.

Atalah, A. (2004). *"The ground movement associated with large diameter pipe bursting in rock conditions and its impact on nearby utilities and structures,"* Bowling Green State University, Bowling Green, Ohio.

Najafi, M. (2005). *Trenchless technology: Pipeline and utility design, construction and renewal.* McGraw-Hill, New York.

National Association of Sewer Service Companies (NASSCO). "International Pipe Bursting Association guidelines," <http://www.nassco.org/pdf/ipbaspec.pdf.> (Accessed December 20, 2005).

North American Society for Trenchless Technology (NASTT). (2004). "Pipe bursting good practices guidelines." NASTT, Arlington, Va.

Trenchless Technology Center (TTC). "Pipe bursting guidelines." *TTC Technical Report No. 2001.02*, <http://www.latech.edu/tech/engr/ttc/publications/guidelines_pb_im_pr/bursting.pdf> (Accessed December 20, 2005).

1.11 DEFINITIONS

Annulus: Free space between the existing pipe and the new pipe.

As-Built: After a pipe has been replaced it is checked for conformance to plan specifications. The actual measurements (as-built) are written on the plan near the original plan measurements. The as-built measurements are often set-off in a box to differentiate the two measurements. Measurements are of the elevations for the manhole top cover, manhole invert(s), ground level, percent grade, and so forth.

Bend Radius: The minimum required arc length of a pipe to safely accomplish a 90-degree turn. The bend radius is approximately 50% larger than the radius of curvature. (See definition of Radius of Curvature and Fig. 1-10.)

Bypass Pumping: The transportation of sewage which flows around a specific sewer pipe or line section or sections via any conduit for the purpose of controlling sewage flows in the specified section or sections without flowing or discharging onto public or private property.

Bypass: An arrangement of pipes and valves whereby the flow may be passed around a hydraulic structure or appurtenance. Also, a temporary setup to route flow around a part of a sewer system.

Carrier Pipe: The pipe that carries the product being transported and that may go through casings at highway and railroad crossings. It may be made of steel, concrete, clay, plastic, DI, or other materials. On occasion, it may be bored directly under highways and railroads.

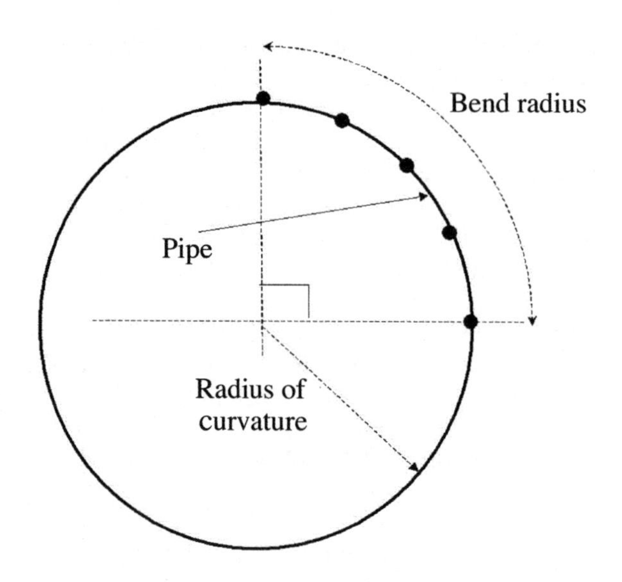

FIGURE 1-10. Pipe bending terminology.

DR: Pipe diameter to thickness ratio.

Existing Pipe: An existing, old, out of capacity, structurally deficient, or deteriorated pipe. In this manual, the term *existing* is collectively used for all of these pipe conditions.

Exit Shaft: See Exit Pit.

Exit Pit: A reception pit where the bursting head can be retrieved.

Force Main: A pressurized pipe that carries sewage under pressure.

Gravity Sewer: A pipe that carries sewage by gravity from a higher point to a lower point.

Heaving: The process in which the ground in front of or above a pipe may be displaced forward and upward, causing a lifting of the ground surface.

Height of Cover (HC): Distance from the crown of a culvert or conduit to the finished road surface, the ground surface, or the base of the rail.

Infiltration: The downward movement of water from the atmosphere into soil or porous rock. The unintended ingress of groundwater into a drainage or sewer system (also termed parasitary flow). Water from the surrounding ground that enters through defects in pipes or joints in a pipeline or through the lateral connections, manholes, or inspection chambers.

Infiltration and Inflow (I&I): The total quantity of water from both infiltration and inflow without distinguishing the source.

Inflow: Flow that enters the sewer. This can be generated by rainfall, an industrial discharge, or another particular connection. Also, water discharged into the sewage system and service connections from sources on the surface.

Insertion Pit: An excavation or manhole into which the replacement pipeline is inserted.

Insertion Shaft: See Insertion Pit.

Lateral: A service line that transports wastewater from individual buildings to a main sewer line.

Launch Shaft: See Insertion Pit.

New Pipe: The new and replaced pipe.

Old Pipe: See Existing Pipe.

Pipe Bursting: A pipe replacement method in which an existing pipe is broken by brittle fracture, using mechanically applied force from within. The pipe fragments are forced into the surrounding ground. At the same time, a new pipe (of the same or larger diameter) is drawn in, replacing the existing pipe.

Pipe Displacement: See Pipe Bursting.

Pipe Eating: A replacement technique usually based on microtunneling, horizontal directional drilling or horizontal auger boring, in which a defective pipe is excavated together with the surrounding soil, as for a new installation. The microtunneling shield machine will usually need some crushing capability to perform effectively. The defective pipe may be filled with grout to improve steering performance. Alternatively, some systems employ a proboscis device to seal the pipe in front of the shield to collect and divert the existing flow, thus allowing a sewer, for example, to remain "live."

Pipe Run: The existing or proposed length of sewer between manholes or inspection shafts.

Pipe Reaming: Removal of an existing pipeline with an HDD and special tooling, where the ground and pulverized pipe are mixed with the drilling fluids and removed with a vacuum truck as the replacement pipe is drawn into place.

Pipeline Rehabilitation: See Pipeline Renewal.

Pipeline Renewal: All trenchless (in situ) methods of pipeline repair, rehabilitation, reconstruction, renovation, and so forth of a pipeline system where a new design life is created.

Pipeline Repair: Reconstruction of short pipe lengths, but not the reconstruction of a whole pipeline. Therefore, a new design life is *not* provided. In contrast, in pipeline renewal, a new design life is provided to existing pipeline system.

Pipeline Replacement: All methods of replacing an existing pipe, where a *new* pipe of the same or larger size is installed.

Pipe Splitting: Replacement method for breaking an existing pipe by longitudinal slitting. At the same time, a new pipe of the same or larger diameter may be drawn in behind the splitting tool. See also Pipe Bursting.

Radius of Curvature: The minimum required distance from the center of the circular path to the arc of the pipe. The radius of curvature is approximately two-thirds of the bend radius. (See definition of Bend Radius and Fig. 1-10.)

Service Lateral: See Lateral.

SDR: See DR.

Upsizing: Any method that increases the cross-sectional area of an existing pipeline by replacing it with a larger-diameter pipe.

PART 2

PLANNING PHASE

The planning of a project has a tremendous impact on the overall project due to the leveraging effect that decisions made at this phase have on all subsequent project activities. Planning is done by all the parties in different stages of construction process, but initial planning is typically done by the project owner using the services of an engineer.

However, the project background may also have a great impact on the planning process itself. The planning phase for a planned and budgeted capital improvement can proceed in a more regulated and orderly fashion than can that for an emergency pipe replacement, such as due to a collapse in a major roadway; the latter is unexpected, unscheduled, unbudgeted, unwanted, and extremely urgent. In an accelerated schedule such as for an emergency, the need for speed of completion may outweigh all other priorities. This may lead to selection of a project delivery system and contracting method that allows a greater focus on speed, such as design-build or cost-plus, rather than preparation of bidding documents along with bid advertisement, opening, review, award, and so forth. These issues should be identified and considered in the first activity (background assessment) of the planning phase.

It is important in the planning process that the project owner clearly identifies, defines, and communicates the project priorities to everyone working on the project. The project owner should also perform a self-audit of its project management to help ensure that these priorities are being followed and that conflicting priorities are not inadvertently introduced into the project performance.

Some suggested planning activities are listed in this section for the planning phase, along with more detailed considerations for the various elements. The suggested planning activities are listed in their most common chronological order which, in general, proceeds from the more general to the more specific concerns. Each activity relies, to some degree, on the previous activity's results. However, as discussed previously, the project background and priorities may require modification, consolidation, rearrangement, or other changes to these activities.

2.1 PLANNING ACTIVITIES

Proper planning helps to ensure that the project will meet the needs and priorities of the owner. The major activities that typically are performed during the planning phase include:

- Establish project requirements and objectives.
- Background assessment: identify risks, constraints, etc.
- Identification and screening of alternatives.
- Data collection.
- Evaluation and selection of alternatives for design.

Further discussion of each activity is presented in the following sections.

2.1.1 Background Assessment

This activity identifies and evaluates the background factors for the project. The background factors are usually nontechnical factors that establish the conditions and limitations of the project, along with influencing the project priorities. Some factors that may be considered for this activity include:

- Is the project a planned improvement or an emergency condition?
- What is the project budget?
- What is the project schedule?
- What are the political considerations (such as if the project has high visibility and/or is in a critical location)?
- What are the impacts and ramifications to public safety, health, or both?
- What are the constraints on construction (by any method)?
- What are the regulatory considerations, impacts, or both?

2.1.2 Screening of Alternatives

This activity will provide a general screening of potential solutions based upon the technical conditions and needs of the project; it will serve to further eliminate candidate solutions and will identify the type of design data that must be collected for the project to proceed. Some factors that may be considerations for this activity (along with subsidiary factors) include:

- Screening for installation of a new pipe, renewal of the existing pipe, or repair of the existing pipe.
  - What are the existing and future land uses for the areas served by this pipe?
  - What are the current and future service requirements for this pipe?
  - What is the ability of this pipe to meet the identified service requirements?
  - What is the condition of this pipe?
  - Is the existing location of this pipe favorable to its continued use in the future?
- Screening for construction methods. Can trenching be used? Can trenchless methods be used? Combinations of both?
  - What are the surface conditions?
  - What are the subsurface conditions?
  - Which alternative is more attractive: installation of a new pipe, renewal of the existing pipe, or repair of the existing pipe?
  - What are the potential geotechnical risk issues?
  - What are the impacts of safety concerns in the project area on the potential construction techniques?
- Screening for methods and materials based upon known, present conditions.
  - What is the slope of the existing pipe? What are the slope accuracy and tolerances needed for the new pipe?
  - What is the diameter of the existing pipe?
  - What should be the diameter of the pipe for the identified service conditions?
  - What are the nature and type of the existing pipe's deficiencies?
  - What effects will the subsurface conditions have on the methods and materials being considered?
  - What are the pipe service conditions?
  - Does the existing pipe have the proper materials for the identified service conditions?
  - What was the method of construction for the existing pipe?
- Screening to identify additional data that may be required to complete this activity or subsequent activities.

2.1.3 Data Collection

This activity includes investigations that provide the information required to evaluate and screen candidate alternatives for consideration in the design phase. Data collection may obtain information relating to:

- Service conditions
  - What are the existing and future service areas and land use?
  - Should flow monitoring be performed?
  - Should system modeling be performed?
  - What are the projections for customer service demands?
  - What is the corrosion potential for the proposed pipe materials?
  - What are the structural requirements?
- Physical conditions
  - What is the topography?
  - What are the surface features?
  - What are the existing utilities and where are they located?
  - What are the sensitive areas within the project site and where are they located?
  - What historical data are available for the project site?
- Subsurface conditions
  - What are the general soil types, locations, and in situ conditions?
  - What is the potential for the presence of rocks, cobbles, or boulders?
  - Is groundwater present? If yes, what is its depth?
  - What is the soil's corrosion potential?
  - What is the soil's settlement potential?
- Existing pipe conditions
  - What is the external condition of the existing pipe?
  - What is the potential for the presence of external voids around the exiting pipe?
  - What are the internal conditions of the pipe?
  - What is the condition of existing manholes on line structures?
  - Where are bends, fittings, valves, service connections, concrete encasements, casing pipes, and other factors specific to the existing pipe located? Does the existing pipe have repair sleeves installed?

2.1.4 Evaluation and Selection of Alternatives

This activity includes the final screening of alternatives and the selection of alternatives for proceeding into design. This final screening is based upon the preliminary known data in addition to the data collected during this activity. Some factors that may be considerations for this activity (along with subsidiary factors) include:

- Determination of the pipe service requirements and conditions
  - What flow capacity is required?
  - What length of pipe is under consideration?

  - What is the corrosion potential?
  - What are the structural requirements?
- Constructability and site limitations
  - What safety issues need to be considered?
  - What type of access into the existing pipe is available?
  - What are the surface impacts of the construction techniques?
  - What are the easement needs of the construction techniques?
  - What impacts will groundwater have?
  - What are the scheduling limitations and constraints?
  - Will bypass pumping and flow control be required during the construction?
  - What is the impact of other utilities?
- Strengths and weaknesses of the potential methods and materials being considered to replace or renew the existing pipe
  - Are the proposed material and method a proven technology with available competent contractors?
  - What is the availability of the technology?
  - What is the anticipated service life?
  - What are the initial and long-term costs?
  - How well does the potential solution satisfy the identified service requirements?
  - What level of quality assurance or quality control is available?

The remainder of this section examines, in more detail, some of these factors and their interrelationships.

2.2 PREDESIGN SURVEYS

Predesign surveys help to identify and obtain preliminary information needed for the design of the project. Although the surveys shown here do not assign the specific party responsible for obtaining the information, it is important to remember that all of this information is needed for project development. It is typically best if the owner and the engineer obtain as much of this information as possible and incorporate it into the design, rather than requiring the contractor to obtain all of it. Information critical to the contractor's successful completion of the project can then be verified by the contractor in the field, prior to beginning the work. The bidding and contract documents should delineate the extent to which the bidders (and ultimately the contractor) can rely on information supplied by the owner.

2.2.1 Land Use

The existing and future land uses in the area affected by the line being replaced are perhaps the most important details that must be addressed in planning capital expenditures. The land use will help to identify the service conditions and service life of the water or sewer line, as well as that of other utilities.

Although this might seem an obvious consideration, it has been overlooked on many projects, usually due to greater consideration being given to shorter-term goals and objectives. However, an existing sewer line that is failing in an area that will undergo dramatic land use changes can either be fixed in a short-term manner or replaced in a way that will accommodate the future use.

This consideration will impact:

- The service conditions of the pipe under consideration
- The design life expectations of the pipe under consideration
- The selection of trenchless versus open cut methods.

Open cut construction almost always requires clearing the right-of-way. If the right-of-way is an alley easement and the clearing allows the municipality to improve the neighborhood starting with the alleys, then open cut may be chosen instead of trenchless methods due to subsidiary benefits other than merely replacing the pipe.

In areas of high-end home values, the land use is extremely unlikely to change and the homes are likely to be maintained for a very long time. In this case, the pipe material selection should be driven by an extremely long design life specification, whether installed by trenching or trenchless techniques.

2.2.2 As-Built Drawings

A review should be made of the as-built drawings and any available documentation from the original contract documents for the line being considered. If the original soil borings are available, they should be reviewed and made part of the supplemental information available to the bidders. The as-built drawings themselves should not be accepted as the final determination of the condition and location of the existing pipe, but should be verified using other surveys described here.

2.2.3 Site Conditions and Surface Survey

Site conditions, including surface improvements, must be thoroughly investigated and documented. The following details should be part of this documentation:

- A thorough mapping of the existing pipe being considered for replacement, including depth and grade as well as the size and material of the existing line. Verify the data shown on the as-built drawings.

- Manhole locations for sewer lines, and valve and fitting locations for water lines.
- Location of other existing utilities, both parallel and perpendicular to the line in question. This detailing must include depth of these other utilities if they are within the potential zone of influence.
- Sewer, gas, and water service connection locations.
- Detailed notes showing surface improvements such as trees and shrubs, type of ground cover, and paving and its condition. All pedestrian and motor vehicle traffic paths must be shown.

When any of the above details cannot be ascertained, the plans should be noted to reflect that the information is missing. All the information must be field-verified by the contractor prior to beginning work.

2.2.4 Subsurface Survey

It must be understood that when designing and specifying trenchless construction, the existing underground site conditions (subsurface) should be investigated and documented to the degree appropriate for the technology selected. The following items may be considered for pipe bursting:

- Excavation of the existing pipe to document the bedding or lack of bedding around the pipe. The location and type of concrete encasements should be documented.
- Excavation of the existing pipe to document the pipe material and joint design.
- In areas of highway crossings, the subsurface survey must ascertain the presence of steel (or any other material) casing pipe or concrete encasement.
- Internal closed circuit television (CCTV) video inspection and field surveys of the pipe location must be part of the early investigation so that the existing grade and suitability for the pipe and replacement methodology can be determined.
- Ground penetration radar (GPR) or other inspection methods may allow an assessment of the potential for voids around the existing pipe. However, if voids are suspected, then a method that allows direct confirmation of voids must be undertaken.
- Internal inspection can help to identify other existing conditions within the existing pipe that can impact replacement, including: partial- or full-diameter collapses of the existing pipe; missing sections of existing pipe; repair couplings; other fittings; offset or misaligned joints; other utilities (such as gas lines) that may have been installed through the pipe; and the presence and location of service connections.

- Sampling and testing of the existing pipe flows to allow a determination of the suitability of the new pipe material and joint gasket chosen.
- A flow test can indicate the condition of water main pipeline interiors. This can be done by measurement of flow and estimating the internal pipe wall friction or roughness.

Some of the different subsurface technologies available simply would not work successfully in certain ground conditions and are not compatible with certain existing pipe materials. Proper information at the planning stage is essential in selecting the proper materials and methodology for the job.

2.2.5 Utility Locating

The predesign survey should locate and identify all of the existing underground utilities in the area within the zone of influence of the existing line being considered for replacement. This can be accomplished by a variety of techniques, each yielding different levels of information and certainty: review of the as-built drawings available from the various utility owners; geographic information system (GIS) data; utility maps; pipeline markers; One-Call location and marking data; and exposing the utility by excavation (usually by vacuum methods) and field surveying its location with respect to the project's coordinate system. As discussed previously, in addition to the clerical effort it may be necessary to visit the site and, in some instances, excavation by an appropriate means will be required.

2.2.6 Subsurface Utility Engineering

This discussion of subsurface utility engineering (SUE) is based upon ASCE Standard, *CI/ASCE 38-02,* "Standard Guidelines for the Collection and Depiction of Existing Subsurface Utility Data" (ASCE 2002). According to these standard guidelines, SUE can be defined as "a branch of engineering practice that involves managing certain risks associated with utility mapping at appropriate quality levels, utility coordination, utility relocation design and coordination, utility condition assessment, communication of utility data to concerned parties, utility relocation cost estimates, implementation of utility accommodation policies, and utility design." The use of SUE offers an opportunity for a more comprehensive and organized approach to locating and accommodating existing underground utilities and can provide more comprehensive information regarding existing utilities that could be present.

Current ASCE standards suggest the use of a "utility quality level which is defined as a professional opinion of the quality and reliability of utility information. Such reliability is determined by the means and methods of the professional. Each of the four existing utility data quality levels is established by different methods of data collection and interpretation." The four quality levels and their associated definitions are (ASCE 2002):

- "Utility Quality Level A—Precise horizontal and vertical location of utilities obtained by the actual exposure (or verification of previously exposed and surveyed utilities) and subsequent measurement of subsurface utilities, usually at a specific point."
- "Utility Quality Level B—Information obtained through the application of appropriate surface geophysical methods to determine the existence and approximate horizontal position of subsurface utilities. Quality level B data should be reproducible by surface geophysics at any point of their depiction."
- "Utility Quality Level C—Information obtained by surveying and plotting visible above-ground utility features and using professional judgment in correlating this information to quality level D information."
- "Utility Quality Level D—Information derived from existing records or oral recollections."

The reader is referred to *CI/ASCE Standard 38-02* for a more in-depth discussion of SUE. However, it is apparent that varying levels of existing utility location information may be applicable, depending upon the specifics of the project. During the planning phase, the owner should decide what quality level of utility information is consistent with the project needs and the owner's risk management strategy, and the potential pitfalls of not having sufficient information. The SUE guidelines offer some potential benefits, including:

- Avoid conflicts with other existing utilities.
- Reduce delays in the construction schedule.
- Eliminate additional construction costs.
- Reduce inconveniences to public.

The quality level of utility information is an important decision that should be made and obtained as early as possible in the project.

2.3. ENVIRONMENTAL IMPACTS AND BENEFITS

As stated previously, the owner and its designer must determine the acceptability of normal construction activity on the neighborhood that is being impacted by construction operations. Construction activities can produce objectionable environmental side effects, such as dust, noise, and vibrations. Dust can be dealt with using water or chemical dust suppressants. Noise control may require additional noise abatement features on equipment, use of smaller equipment, erection of sound barriers, scheduling of work to avoid disturbing noises being heard, or other measures.

Lane closures and traffic patterns can be dealt with as they are in other kinds of construction. The question of vibration has come up because a great deal of work has been done with percussion hammers. A review of the methodologies listed here will show that the use of percussion hammers is only one of many available methods.

The owner must have a clear understanding and a realistic expectation concerning the material and method that has been chosen to replace the existing piping in its local community. The use of pipe bursting can yield significant benefits and lessen concerns with regard to environmental impacts due to a number of factors, including limited work locations during the project, less use of heavy equipment, less equipment movement required along the project, less material usage such as embedment, and the actual location of the pipe replacement being buried during the replacement operation.

2.4 SOCIAL IMPACTS AND BENEFITS

Trenchless replacement of underground infrastructure normally has social benefits as follows:

- Reduces pavement repair cost.
- Reduces the possibility of damage to surrounding utilities, structures and pavement.
- Reduces vehicular traffic disruptions (however, this can be negated by extensive barricading to accommodate fusing a long string of pipe prior to beginning work).
- Enhances safety by reducing the amount, size, and duration of excavations.

2.5 PERMITS

Permits are typically permissions granted to the utility owner from the owning entity of an existing facility, to construct or replace a pipeline under the existing facility. Permits typically have construction requirements and conditions for continued occupation of the space beneath the existing facility (such as annual fees, maintenance requirements, reporting requirements, and insurance). Although the permitting entity will hold the utility owner ultimately responsible for any violations, the contractor will also be responsible (to the utility owner) for compliance with permit requirements pertaining to the construction. Permits can require a lengthy processing time to obtain and therefore are typically obtained by the utility owner during the project design phase. The contract documents should list and contain a copy of the construction requirements for all permits obtained for the project. The contract should also require the contractor to adhere to the requirements of the permits.

Some typical locations and types of permits that could be required for a given buried utility project include:

- U.S. Army Corps of Engineers 404 Discharge Permit
- Wetlands crossing permits
- Floodplain development permits
- Crossing permits for:
  - City streets
  - County roads
  - Federal and state highways
  - Interstate highways
  - Railroads
  - Waterways
- Construction permits can be issued by:
  - Municipalities
  - River authorities
  - Counties
  - Regional agencies
  - State and federal regulatory agencies
  - Funding agencies.

It is very important that the required permits be identified as early as possible during the project planning phase and measures be instituted to secure the permits in a timely fashion.

2.6 JOBSITE LOGISTICS REQUIREMENTS

The owner must recognize and deal with community and neighborhood needs in setting up the following details of the contract documents, which will direct the contractor in its work:

- Traffic control
- Storage areas
- Equipment setup areas

- Construction staging areas
- Location of major supporting equipment
- Dust and noise restrictions
- Allowable working hours and working days
- Blackout periods, if any.

This information should be compiled in the contract requirements to avoid violation of private properties or unreasonable interference with any public or private operations. When using pipe material that must be butt-fused together into long lengths, a barricade plan should be part of the planning process. During the planning phase of the project it is important to recognize the logistical space requirements associated with the anticipated construction techniques. In some instances, the availability of space (or lack of it) for logistics will be the controlling factor in selecting construction techniques.

2.7 LENGTH OF INSTALLATION

The length of the line segment that can be replaced is another detail that will be impacted by the material and method that are chosen. When segmental rigid pipe is chosen, the pit location can be placed wherever it will have the least impact on the area and on costs. If the manholes are being replaced, the pits will be located at the manhole locations and the push can be received into an excavation at service lateral—wherever it is easy and convenient. If the manholes are being saved, the pits could be located at a service lateral excavation and lesser manholes could be used as receiving pits.

When flexible pipe is chosen, the impact of fusing the new pipe together before construction can begin will dictate pit locations. Usually, the length of the sewer burst is determined by distance between manholes. For sewer pipe, this distance is typically around 300 to 400 ft (90 to 120 m).

With all pipe bursting methods and materials, the bursting head may pass through a manhole, which can be termed an *intermediate manhole*. The intermediate manhole must be prepared to allow the equipment to pass without offering any resistance. This will normally necessitate enlarging the wall opening and removing the benching to prevent the hammer or expander from rising in the manhole.

With some pipe bursting materials and methods, lubricant must be used to reduce pulling forces in the segment being worked. This will impact the material chosen. The owner must recognize the limitation of tensile strength of any selected pipe system (e.g., pipe material and joints).

In addition, possible replacement distances are dependent upon the pipe weight, frictional (drag) characteristics, installation method, and response of the surrounding soil on the pipe outer surface. There is presently no methodology accepted industry-wide for quantifying these effects. However, based upon limited experimental field data, there is evidence that corresponding induced tensile forces can vary over a wide range. This implies that it may be difficult to confidently predict practical placement distances for a given pipe system (e.g., specified material, diameter, wall thickness), thereby complicating any proposed design process.

2.8 ACCURACY AND TOLERANCES, INCLUDING SETTLEMENT AND HEAVE

As with other details of the planning phase, the amount of settlement and heave will be greatly impacted by the choice of new pipe material and the method chosen. The latest developments in pipe bursting technology have allowed replacement work to be successfully performed in tougher ground conditions.

The designer would be well advised to fully understand the technology that has been chosen and have expectations that are based on construction reality. This can be very difficult because the existing ground conditions can change dramatically from one area to another, especially given that the work is being done in ground conditions that may or may not include imported material. In addition to the existing material, the moisture content can impact the ability of the ground to be compacted around the new pipe. This further emphasizes the need for obtaining good subsurface information during the planning phase.

It should be remembered that the new pipe has a specific wall thickness and the bursting pathway that is created must include an annular space to allow the new pipe to be pulled through it. Even when the line is being replaced size-on-size, there is an upsizing going on because the expander must be larger than the new pipe OD and because the initial bore diameter is the existing pipe ID.

If the project is near a structure or under the pavement, and the public agency (such as municipality, state department of transportation, etc.) and the contractor need proof (after the fact) that the trenchless pipe replacement operation did not affect the structure or the road embankment, then monitoring should be undertaken. If the line is across a golf course, the costs of monitoring are hard to balance against the benefits. Sometimes the cost of repairing heave over the pipe being replaced is less than the cost of open cut construction. Where heave is expected to occur, it is easy to saw cut the pavement over an existing line before the work begins. After the line has been replaced, the saw cut heaved portion of the pavement can then be replaced. This approach may be less expensive than open cut construction, which could impact local business and traffic in a much more disruptive way. On the question of settlement, there are ground conditions where the use of percussion will result in consolidation of the ground, with possible settlement of the ground.

2.9 INSERTION AND RECEPTION PITS

Working pit locations (both insertion and receiving) will be affected by the pipe material and methodology chosen to perform the work. In this manual, insertion pits will refer to the pits where the pipe bursting head enters the existing pipe. Receiving pits will refer to pits where the pipe bursting head exits the existing (host) pipe after being pulled or pushed through and bursting the existing pipe.

An additional consideration is the need to replace manholes, valves, or fittings which will be greatly impacted by the pits. When planning the required pit locations, it is useful to first determine where there will be excavation of one kind or another independent of the materials or methods that are being employed. If the manholes, valves, or fittings are being replaced anyway, the necessary excavation can be an insertion or receiving pit location. If the manholes are to be saved, any location where excavation is necessary (e.g., service laterals) can be used.

If the specifications call for a segmental pipe product, one of the following methods may be used:

- Jacking—this methodology is similar to pipe jacking. The insertion pit must accommodate the jacking frame, one pipe joint, and other facilities similar to the jacking pit for a pipe jacking installation. It requires a receiving pit that only needs to be large enough for removal of the pipe bursting head, and usually can be as small as an existing manhole.
- Pulling—this methodology pulls the rigid pipe by using a tow cable or pull rod that is run through the new pipe and connected to a backing plate on the trailing end of the new pipe string. This is similar to a pipe bursting head that is being pulled through the existing pipe by the tow cable or pull rod; however, the new pipe is actually being pushed by the backing plate and each subsequent pipe section in the string of new pipe. For static pipe bursting methods, the reception pit must be large enough to allow for removal and installation of the backing plate on the tow cable or pull rod and installation of a pipe joint in the pull string. For pneumatic pipe bursting methods,

the reception pit must be large enough to allow for installation and operation of the winch or pullback device, along with removal of the pipe bursting head.

When the specifications call for a flexible pipe product, the following considerations are appropriate:

- The insertion pit must be long enough to provide a gently sloping ramp for the new pipe into the pipe burst hole. Manufacturers' bend radius (pipe minimum radius of curvature) guidelines for insertion pits require that the pipe must enter the ground gradually so as to preclude damaging it as it enters the existing pipe. The Plastic Pipe Institute (PPI) provides such guidelines for PE pipes.
- The receiving pit must be large enough to allow for installation and operation of the winch or pullback device, along with removal of the pipe bursting head.

In constructing and maintaining the insertion and receiving pits, the contractor should consider the following:

- The goal of maximizing the length of the new pipe installed at one time might seem like a positive direction for planning. However, in the field during the day-to-day work, the job may be better managed by shortening the line segment being worked, to reduce risks.
- Providing a survey of the various utility companies' facilities would minimize possibility of damage to these utilities.
- The work should be planned and carried out with respect for street rights-of-way and other utility easements.
- Contractors are generally held responsible for repairing any damage to the property of other parties as a result of the pipe replacement process.
- Contractors must provide and install trench shoring or bracing in compliance with Occupational Safety & Health Administration (OSHA) standards.

2.10 SERVICE CONNECTIONS

Service connections can be treated in a number of different ways, depending on:

- Who owns the service lateral
- Who owns the service connection at the main

- The limits of property ownership, the utility easement, or both
- The traditions and procedures of the local agency.

Usually, the connection of the service line to the main must be approved by the utility that owns the main line. In some instances, the owning utility may make this tap with its own crews, whereas in other locations a licensed plumber must be hired by the property owner to make the tap, which is then inspected by the owning utility. A myriad of potential arrangements exists in this area, depending upon location.

The planning phase is the best time to clarify the responsibilities of who will disconnect, reconnect, test, inspect, and approve service connections and their associated laterals. Although it is risky to generalize about this, past experience suggests that if the lateral is owned by a private property owner, the following rules apply:

- The connection at the main will probably be done under the authority of the municipality, and therefore the contractor replacing the main line will reconnect the lateral.
- Excavating near the main prior to the replacement process, no matter the material or method, is *not* recommended. This is because the differing compaction of the excavated soil could allow for a deviation from line or grade in the pipe bursting head.

If the lateral is owned by the municipality, the following rules apply:

- The line should be replaced to the edge of the right-of-way and consideration should be given to installing a new cleanout at the property line.
- The considerations of design life should apply to the saddle configuration as well as to the material used to replace the lateral by open cut.

All house leads or laterals must be located prior to starting the work. When the main line is being replaced by a radial force that is expanding the existing pipe, there is a possibility that the radial force could act upon the column of existing lateral pipe and cause it to move along its axis. This phenomenon could result in damaged pipe. If the lateral is exposed and severed at an appropriate location, the break can be controlled.

2.11 MAINTAINING SERVICE

The maintenance of service to the customer should be planned so as to minimize customer inconvenience, ensuring that the public will not experience any health risk or lapse of service. This is important regardless of whether the line is water, sewer, or any other utility.

2.11.1 Water

Plans must identify the line segments that will be replaced.

- When a line segment is cut at an existing isolation valve, the valve must have appropriate thrust blocking in place.
- All manifolds that will carry fresh water to the customer must be chlorinated and tested before going into service.
- All manifolds must be protected against traffic as well as vandalism.
- All of the property owners must be made aware that the manifolds will cross their property and may cross their driveway(s).

2.11.2 Sewer

- When planning for bypass pumping of sewage, there may be more lines affected than just the line segment being replaced.
- The only way to shut off the flow of sewage is to shut off the flow of water to the area upstream, and that is very seldom tolerated.
- During the planning phase, the sewage generated by each property must be addressed as to its volume and the ability to intercept the sewage at the lateral at the property line.
- In order to cross traffic arteries, the bypass line may have to run through storm sewer conduits and they may not be adjacent to the line segments being worked.
- The bypass plan must include any necessary permitting.

2.12 COST CONSIDERATIONS

Construction project costs are based on direct costs (labor, materials, and equipment), indirect costs (field overhead and home office overhead), and markup (insurance, bonding, contingency, and profit). Because trenchless technology methods are relatively new, historical costs may not be as available as they are for conventional open cut methods. In addition, many mid-sized or small communities may have a hard time finding qualified and experienced contractors to submit bids for trenchless technology and, specifically, pipe bursting projects. Engineers and utility owners need to evaluate the specific conditions of each project and select the method or methods that can safely and cost-effectively fulfill specific

TABLE 2-1. Guidelines for Selection of Pneumatic or Static Pipe Bursting Methods

Existing Pipe	Pneumatic	Static
Metallic pipes including aluminum, copper, ductile iron,[a] wrought iron, steel, or stainless steel	No	Yes
Plastic pipe, including HDPE or MDPE,[a] PVC, CIPP, or fiberglass	Yes	Yes
Prestressed or bar-wrapped concrete cylinder pipe (PCCP or BSCCP), corrugated metal pipe (CMP), or corrugated plastic pipe	No	No
Fracturable pipes, including asbestos cement (AC), RCP or nonreinforced concrete pipe, CI, VCP, or Orangeburg	Yes	Yes
Valves, stainless steel clamps or repair bands, point repairs[b]	No	No
Pulling in prechlorinated replacement pipe	No	Yes

[a]Limited distances, depending on upsize and existing conditions of surrounding soil.
[b]Limited success with static system only; recommend open cut at these locations.

project objectives and requirements. The method selection option should not be left to contractors because contractors are generally optimistic and may want to execute a project for a number of reasons, such as a current lack of work and availability of equipment and crews.

Another important consideration is the social costs of utility construction, which are not included in this cost breakdown. Social costs include inconvenience to the public; traffic disruption costs; damage to pavement, existing utilities, and nearby structures; noise and dust problems; loss of business; safety hazards of trenching; and damage to the environment and green areas. Trenchless technology methods (specifically, pipe bursting) will reduce social costs and, if these costs are considered, trenchless projects prove to be more cost-effective than open cut methods. However, contractors cannot include these costs in their bids; social costs can only be considered by owners and engineers within a life-cycle-cost approach to project planning.

Project-specific conditions will be cost drivers for all construction projects. In all instances, consulting and partnering with experienced and reputable contractors can assist in determining potential project pricing.

2.3 GENERAL SELECTION GUIDELINES

General guidelines for selection of a pneumatic or a static pipe bursting method (the most common pipe bursting methods) for different existing pipe materials are presented in Table 2-1.

PART 3

EXISTING (HOST) PIPE

As mentioned previously, pipe bursting/replacement is a trenchless or semitrenchless pipeline installation technique for recovering lost capacity and structural integrity of an existing pipe. The structural integrity is achieved by replacing a deteriorated piping system or (for increasing capacity) by replacing an undersized piping system. One goal is to accomplish this task with minimal surface excavation. Utilizing the existing (host) pipe as a pathway, most pipe bursting/replacement systems utilize a process whereby the existing pipe is mechanically fractured or split, expanded, and displaced radially outward into the adjacent soil. Displacement of the fragments of the existing pipe is generally accomplished with a cone-shaped expansion (or bursting) head forced through the relatively smaller-diameter original pipe by means of hydraulic, pneumatic (percussion), or static axial force from a pulling or pushing mechanism.

Most existing pipe materials used for water, wastewater, and natural gas piping systems are candidates for bursting using present replacement methodologies. There are, however, situations where an existing pipeline may not be a good candidate for bursting due to the soil environment around the pipe, such as pipe installed in a rock trench, backfilled with concrete or certain other fill encasements, or a pipeline at a shallow depth. In addition, there are a number of pipe materials for which technology is not presently available for fracturing, splitting, or otherwise displacing the burst pipe or its fragments into the adjacent soil. Thus, the first step in evaluating the compatibility of a given piping system using pipe bursting/replacement is to address the issues related to the existing pipe and its soil environment.

3.1 GENERAL CONDITION ASSESSMENT

For the existing pipe, the type, material, and associated components (i.e., reinforcement, joint type) have a direct impact on the viability of the bursting/replacement methodology and the design of the expansion cone. Various pipe materials used in both water and wastewater service, and their potential for being burst and replaced using these trenchless technologies, are discussed in detail in Section 3.3.

Once it has been determined that the existing pipe is a candidate for bursting technology, the structural condition, alignment, service connections, and appurtenances must be assessed. Condition assessment generally results in existing piping systems falling into one of two groups: (1) existing piping systems that are in a serviceable condition but are, from a hydraulic perspective, undersized and require upsizing, or (2) piping systems that have some degree of deterioration that inhibits flow, dependability, or both.

3.1.1 Undersized Pipe

Most owners of water and wastewater utilities have hydraulic modeling capabilities for their systems. This is helpful in quickly providing feedback related to the requirements for flow through a given portion of a pipe system. Many of these same modeling programs also maintain information about the size and type of pipes, location, and frequency of repair for the system. Locations are often stored in the owner's geographic information systems (GIS) that can provide nearly exact locations of pipelines, appurtenances, point repairs, and so forth. These owners have the luxury of being able to accurately store information about the locations and alignment of existing facilities and infrastructure as well as import location data from other piping systems.

After the pipeline has been hydraulically analyzed and the conclusion has been reached that a larger line would be needed, consideration of the capabilities of the pipe bursting equipment must now be addressed. The feasibility of upsizing depends on a number of factors, including soil conditions, trench geometry, diameter increase, volume increase, and bursting length (North American Society for Trenchless Technology (NASTT), 2004). Generally, pipe bursting/replacement equipment is capable of increasing the diameter of the new pipe by two nominal sizes (e.g., existing 6-in. pipe may be enlarged to either 8-in. or 10-in. pipe). Although increases larger than two nominal sizes have been successfully completed, the pipe bursting guidelines of the National Association of Sewer Service Companies (NASSCO) consider increases of three nominal sizes to be in the most difficult category. The NASSCO Project Design Classification is provided in Fig. 3-1, designating the three categories of difficulty (A, B, C), with Case A being the least difficult and Case C the most difficult. Appli-

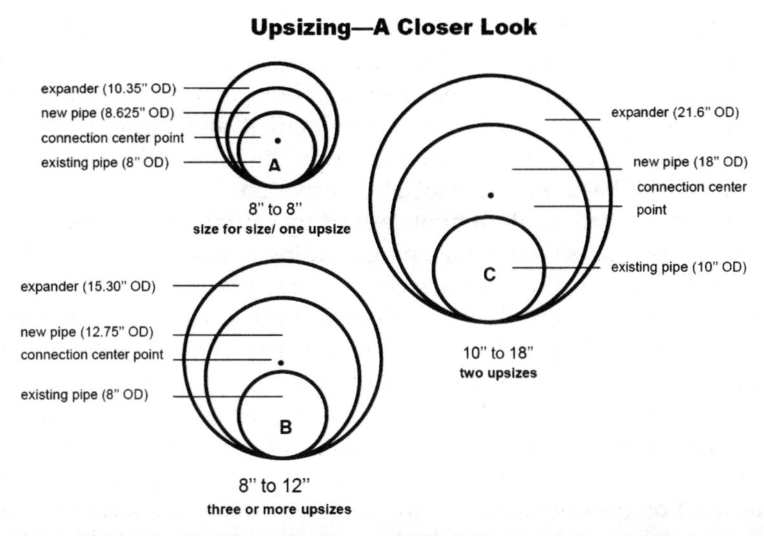

Case	Depth of Pipe	Existing Pipe Diameter	New Pipe Diameter Options	Burst Length
Case A	<12 ft	4" - 12"	Size-for-Size or one (1) nominal upsize	0 - 350 ft
Case B	>12 ft <18 ft	12" - 20"	2 nominal upsizes	350 ft - 450 ft
Case C	>18 ft	20" - 36"	3 or more upsizes	> 450 ft

FIGURE 3-1. NASSCO upsizing guidelines.

cations corresponding to great depth, long burst length, or large diameter also increase the level of difficulty of the pipe bursting operation.

3.1.2 Deteriorated Pipe

Utilities generally have established criteria for determining when a portion of a piping system is deteriorated to the extent that it requires replacement. It is of paramount importance that the suspected deteriorated line be inspected (via CCTV or, alternatively, scanned) to determine the extent of any blockage or structural deficiencies. In addition, statistical information on the frequency of breaks and repairs and the projected cost for maintaining the line based upon the present value of projected future costs would support the decision for replacement.

Pipelines that are designated as structurally deficient include pipe sections with evidence of existing failure, pending failure, or both. In gravity applications, rigid pipes (such as concrete or clay) typically indicate distress from external loads by the relative severity of cracking at the crown or invert on the interior of the pipe. Flexible (low stiffness) pipes typically indicate distress by exceeding allowable deflection or actual collapse (buckling) at the pipe invert. In very aggressive sewer environments (pH = 1 to 4), pipe structural deficiency is likely to be the result of internal corrosion. Sewer mains containing an anaerobic environment (e.g., formation of slimes with no free oxygen, prolonged air spaces, high temperatures, long detention times, low velocities, and formation of high hydrogen sulfide gas concentrations) inside the pipe will likely result in acid-producing bacteria and associated corrosion.

Changes in vertical alignment caused by initial misalignment, poor installation practices (inadequate pipe bedding and backfilling), or differential settlement or flotation are characterized by misaligned joints or swags that may adversely affect the hydraulic performance of gravity pipeline systems. Depending upon the severity of the misalignment or swag, static pipe bursting systems, applied to existing pipes with substantial beam strength (e.g., DIP or VCP) may help to straighten minor alignment variations. If the alignment variations are excessive or abrupt, it is best practice for the owner and engineer to plan for appropriate point repair(s). Longer sags (more than 10 ft) will likely remain in the new replacement line and result in higher maintenance, potential overflows from siltation, loss of capacity, and an increase in hydrogen sulfides. The severity and impact of remaining long sags should be considered by the owner and engineer during the overall project evaluation.

3.2 SPECIAL CONSIDERATIONS

In general, owners and engineers should avoid stating requirements for means and methods for construction operations. Installation specifications should be performance-based. It is, however, important to identify potential obstacles which may limit the applicability of some of the bursting techniques so that appropriate information is available for the purpose of soliciting project bids where the risk to the bidding contractors has been minimized.

It is essential to have information regarding the type and condition of the existing pipe and appurtenances. If the pipeline is being replaced because it is undersized (but is otherwise in good physical condition) and the bursting cone is appropriately designed, the pipe will tend to resist (without collapse) the axial component of the force applied during the bursting operation. If, however, the pipe is being replaced because of deterioration (e.g., internal corrosion, external corrosion, or both) and the deterioration is not uniform through the wall of the pipe, it is possible that the axial force component being transmitted through the pipe wall may collapse and longitudinally telescope the existing pipe. This effect may result in the thinner material accumulating into a thicker, irregularly shaped mass in front of the bursting cone, possibly preventing a successful bursting operation.

Potentially liquefiable soils in wet environments require additional study. Loose, granular, saturated soils have the potential to liquefy from vibrations associated with pipe bursting systems that use percussion to advance the bursting head. This could possibly result in induced dramatic grade deviations, even though the original line had an acceptable grade and alignment. Proper identification of these soil characteristics is necessary to ensure that the most appropriate bursting equipment and materials are used.

In limited construction site space or unusually dense commercial or residential areas, the site restrictions that may impact and limit equipment use, the new product pipe material, and the installation methodology (cartridge or assembled line) should be defined. Access to the site, maximum allowable work area, traffic flows to businesses and residences, restrictions on work hours, and limits on the amount of vibration on adjacent structures are important factors to consider. Considerations for these and other restrictions should be addressed by the owner and engineer during the planning stage.

The most important difference between existing gravity service pipe systems and existing pressure service pipe systems is related to the degree of alignment. This is an important issue when considering pipe bursting/replacement systems that use direct jacking force applied directly to the new pipe to advance the bursting head. In gravity service installations, the vertical and horizontal alignment only changes at manholes, with the intermediate path installed in a relatively straight line. Therefore, direct jacking loads applied to a replacement pipe are more uniformly distributed around the circumference of the pipe's cross section. In contrast, pressure pipelines vary the vertical alignment as necessary to maintain a minimum cover and change horizontal alignment as required, without regard to the degree of straightness or alignment between access points. Direct jacking of the associated replacement pipe, if not fully engaged, places a concentration of stresses that may compromise the safety factor built into the joint. Other specific issues are discussed in the following sections.

3.2.1 In-Line Joints and Appurtenances

As-built drawings, previous inspection and maintenance records, and the knowledge of maintenance personnel about the existing pipe must be researched. In-line appurtenances such as valves in water and sewer force mains, bends, manholes, tees and wyes, and other potential in-line obstructions must be field-verified.

Although bell and spigot steel rings used for joining various pipe types are only occasionally used in gravity pipelines, they are commonly employed in pressure applications. Metallic stiffening rings of various types are also used when connecting PE pipe and appurtenances, creating mechanical joints that rely upon compression of the gasket against the pipe for sealing purposes. These rings may be problematic if the force necessary to split or fracture them is greater than the ability of the soil friction to maintain the axial position of the pipe. When this occurs, it is possible for entire sections of pipe to be pulled or pushed back to the equipment or exit pit. Alternatively, if these rings are not properly slit or expanded, they may collect on the front of the head, resulting in significantly greater resistance or drag, or possibly halting the bursting/replacement operation.

3.2.2 Location and Description of Point Repairs

Due to the variation in capabilities of present-day pipe bursting/replacement equipment, the owner and engineer must identify the locations of all point repairs made to the pipeline section that is being studied for replacement. At each location, the type of repair must be described as accurately as possible and include, as a minimum, the information found in Table 3-1.

The pipe bursting/replacement equipment may be capable of breaking up the existing pipe but not capable of breaking stainless steel repair sleeves, replacement sections of ductile iron, steel pipe, or concrete or reinforced concrete encasement without special cutting heads or other special equipment.

TABLE 3-1. Required Information for Each Previous Point Repair Location

Type	Information Required
Type and size of affected pipe	Type, Diameter
Adjacent pipe condition	Describe
Cause of leak or damage	Describe
Length of repair or replacement	Length
Clamp required, type and size	Type, Diameter, Length
Solid sleeve required, type (CI or DI), and size	Type, Diameter, Length
Pipe required, type and size	Type, Diameter, Length
Solid sleeve required for pipe, type (CI or DI), and size	Type, Diameter, Length
Backfill material	Concrete, Stone, Flowable Fill, Other

It is extremely important to assess the general suitability of the existing pipe to be fractured (or split) and radially displaced outward into the adjacent soil. Pipe materials have different mechanical and performance characteristics that make a particular material appropriate for sewer, water, or other applications; few pipe materials are viable for all functional categories. This section discusses the ability of various pipe materials to be burst/replaced using presently available methods and equipment, recognizing that new technologies will be developed that will overcome current limitations. Pipe materials are therefore divided into three classifications: (1) Fracture and Expand, (2) Split and Expand, and (3) Limited or No Existing Technology, as described in Table 3-2. In addi-

TABLE 3-2. Suitability of Existing Pipe Material for Bursting/Replacement

Pipe Type	Fracture and Expand	Split and Expand	Limited or No Existing Technology	New Replacement
Asbestos cement pipe (ACP)	X	—	—	X
Bar-wrapped concrete cylinder pipe	—	—	X	N.A.
Concrete pipe—reinforced (RCP)	X	—	—	X
Concrete pipe—nonreinforced (CP)	X	—	—	X
Concrete pipe—polymer (PCP)	X	—	—	X
Ductile iron pipe (DIP)	—	X	—	X
Fiberglass pipe (FRP, GRP, RPMP)	—	X	—	X
Gray cast iron pipe (CIP)	X	—	—	N.A.
High-density polyethylene (HDPE)	—	X	—	X
Polyvinyl chloride pipe (PVC)	X	X	—	X
Prestressed concrete cylinder pipe (PCCP)	—	—	X	N.A.
Vitrified clay pipe (VCP)	X	—	—	X

tion, pipe materials that may be used as a new replacement pipe installed by the pipe bursting procedure are identified in the last column.

3.3 FRACTURE AND EXPAND

Existing pipe materials are considered fracturable if the pipe experiences brittle catastrophic failure when subjected to a radial expanding (or tensile) force. In general, fracture and expand pipe materials have mechanical properties which are either very low in tensile yield strengths or have very low elongation characteristics (they are brittle). Pipe materials with these properties are good candidates for pipe bursting. They include ACP, CP, RCP, PCP, VCP, and CIP.

3.3.1 Asbestos Cement Pipe (ACP)

Asbestos cement pipe was widely used in water and, to a lesser degree, in sewer applications in the United States until a controversial U.S. Environmental Protection Agency (EPA) ban in 1989 and a corresponding phase-out plan by 1996. Although the formal ban was lifted in the early 1990s, the phase-out plan has been quite successfully implemented. However, ACP continues to be a prevalent pipe material in many parts of the world today. Designed as a rigid conduit, the class and corresponding wall thickness of the pipe are determined by the combination of internal pressures and external loads. Although ACP is structurally a good candidate for pipe bursting/replacement, the owner and engineer must investigate any federal, state, or local regulations that might prohibit abandoning this pipe material in situ after it is fractured. Some regulations may consider burst asbestos pipe a potential hazard, even when left underground and outside the new pipe.

3.3.2 Concrete Pipe (CP)

Concrete pipe is designed as a rigid conduit where the external earth loads are designed to be transferred through the pipe wall into the soil beneath the pipe. There is a wide variety of concrete pipes available for both pressure and gravity service, including CP (also known as Packerhead pipe), RCP (C76), PCCP, RCCP, bar-wrapped steel cylinder, and PCP.

Concrete material has a relatively high compressive strength but an inherently low tensile strength (i.e., its strength in tension is only about 10% of its compressive strength). Thus, standard CP is ideally suited for the pipe bursting/replacement process, during which the bursting cone generates tensile stresses within the walls of the pipe, causing fracture of the existing pipe.

RCP and other pressure cylinder pipes, however, incorporate a steel reinforcing cage or solid steel cylinders on the inside of the pipe for addressing loading conditions requiring a significant increase in tensile capability (e.g., relatively large pipe diameter, increases in burial depth, or both). This steel reinforcing cage restricts the number of bursting systems that can be used for replacing this type of pipe.

PCP was introduced into the United States circa 1997. Polymer concrete uses thermosetting polyester resins and select aggregate only. Mechanical properties for polymer concrete exceed those for Portland cement concrete. However, similar to standard nonreinforced pipe, PCP can be burst/replaced due to its relatively low tensile strength and limited ductility.

3.3.3 Gray Cast Iron Pipe (CIP)

With the exception of gray cast iron soil pipe for architectural applications, CIP is no longer available for use in pressure pipes. CIP, the predecessor of modern-day DIP, was designed as a rigid conduit due to the material's lack of flexibility. This low degree of flexibility allows the relatively brittle CIP to be effectively burst and replaced. CIP typically breaks off in slabs which, in yielding soil trenches where the fragments can be properly expanded, do not normally tend to damage the new plastic pipe during the pull-in replacement process (additional discussion in Section 4.4.1.1). Scratching or gouging is generally not a concern for long-term performance of pipes other than plastic pipes (e.g., HDPE and PVC).

3.3.4 Vitrified Clay Pipe (VCP)

Vitrified clay pipe is one of the most inherently inert pipe materials available (i.e., it is resistant to a broader range of pH values and contaminants than any other pipe material). Because it displays excellent compression strength and poor tension characteristics, VCP is also designed as a rigid conduit. The manufacturing process for VCP prohibits the use of secondary steel reinforcement which, in combination with the relatively low tensile strength and lack of ductility of this ceramic pipe, renders this material the ideal candidate for pipe bursting/replacement. Similar to CIP, VCP generally breaks off in slabs which, in yielding soil trenches where the existing pipe fragments can be properly expanded, do not normally tend to damage new plastic pipes during the replacement process. Once again, scratching or gouging is generally not a concern for long-term performance of pipes other than plastic pipes (e.g., HDPE and PVC).

3.4 SPLIT AND EXPAND

Existing pipe materials that are not considered to be brittle have mechanical properties consistent with either high tensile strength or moderate ductility (as measured by the material percent elongation during a tensile test), or relatively low tensile strengths with extreme high ductility (elongation). Such materials are unlikely to be expanded sufficiently to allow the required clearance for the trailing replacement pipe. Systems have therefore been developed that operate in two stages. In the first stage, the pipe is slit, splitting the pipe longitudinally. The required tool is typically configured with a series of successively larger hardened disk cutters, aligned along the cutter's long axis; hydraulically activated cutting blades; or hardened or carbide-tipped cutting wings. During the second stage, a cone-shaped expanding head, either integral to the splitting tool or immediately behind it, forcibly expands the existing pipe into the adjacent soil to generate adequate clearance for the trailing replacement pipe.

Replacement of DIP or steel pipes, using the process of split and expand, requires a substantial amount of axial thrust to expand the slit pipes. In general, this method for pipe replacement is limited to static pull or push systems.

3.4.1 Metallic Pipe

Existing metallic pipes for sewer applications will generally include DIP and, in some systems, steel pipe, both smooth-wall and corrugated. Both materials have approximately the same mechanical properties. With a minimum tensile strength of 60,000 psi and a minimum elongation of 10%, metallic pipes are clearly some of the most difficult to burst/replaced. Furthermore, there is the possibility that the axial slit/split, cut into the wall of either DIP or steel pipe prior to the expansion, may present sharp edges to the replacement pipe. When attempting a large upsizing, this condition may detract from the long-term performance of plastic pipe due to external scouring, cutting, or gouging of the pipe wall. Such external damage would have a much greater impact on plastic pressure pipe than on plastic gravity service piping applications. In order to reduce such occurrences, the pipe burst operation should be limited to size-on-size or to an increase of only one nominal size.

Similarly, when bursting existing pipe installed in an environment that has been determined to be corrosive to metallic pipes, any new steel replacement pipe with an exterior bonded coating system for corrosion protection can be compromised by scoring, cutting, or gouging of the protective coating. However, there are corrosion protection methods available which are appropriate for some types of metallic pipe installed by the pipe bursting method. These methods may require that the pipeline be either welded or, for rubber-gasketed joints, utilize some type of joint bonding to provide electrical continuity. This continuity allows the pipeline to be monitored for the development of active corrosion cells. At any time during the life of the pipeline, as indicated by changes in the pipe–soil potential, life-extending cathodic protection can be added (related discussion in Section 4.3.2.).

3.4.2 Plastic Pipe

Plastic pipes of interest include fiberglass, HDPE, and PVC. Filament-wound fiberglass pipe can have tensile strengths on the same order of magnitude as metallic pipes. Therefore, these pipes typically require the process of slitting/splitting prior to expanding. However, random-oriented fiberglass pipe, such as reinforced plastic mortar pipe, may be burst by expansion. Due to their high ductility characteristics, existing HDPE and PVC pipes also require slitting/splitting prior to replacement.

3.4.3 Limited or No Existing Technology

Two types of concrete pressure pipe manufactured using multiple components are not economically feasible to burst using existing technologies. These include prestressed concrete pipe (both with and without a steel cylinder) and bar-wrapped concrete cylinder pipe. Historically, the smallest-sized pipes for these materials have been 16-in. and 12-in., respectively, with the basic structure of the pipe being concrete (with or without steel cylinders) with secondary wraps of either high-strength prestressing wire or mild steel rod. It is this combination of a steel cylinder with wire or rod that prevents present technologies from successfully bursting/replacement such pipes.

PART 4

NEW (REPLACEMENT) PIPE

There are several bursting systems available in the industry for application to different types of pipe materials. While the existing pipe is being fractured or split and expanded radially outward into the adjacent soil, the new replacement pipe is simultaneously installed. Pipe bursting/replacement systems use one of three basic methods for installing the new pipe behind the expansion head. The majority of these methods apply a sustained tension or pulling force to the bursting head via oil-field type solid pulling rods, flexible chains, or cables (wire rope) fed through the existing pipe from the pulling machine at the receiving pit or manhole. The new pipe is mechanically connected to the expansion head and is pulled into the cavity within the expanded existing pipe.

Although proper pipe sizing and economics are important considerations during the planning, design, and selection of the new pipe system, there are several other issues that should be addressed. These include:

- Fluid being transported (water, wastewater, other).
- Availability of diameter sizes, pipe section lengths, and joints for use with the particular trenchless technology methods.
- Actual ID or flow diameter. For a given nominal size, pipes of various materials with the same pressure rating do not necessarily have the same ID, possibly requiring additional upsizing to achieve the proper hydraulic design.
- Volumetric upsizing capability is determined by the largest possible replacement OD that may be inserted and the actual ID of the original pipe.
- Internal surface flow coefficient as based upon in-service field experience—not just theoretical values obtained from laboratory testing.

- Construction and operational stresses. For trenchless technology projects, the stresses induced in the pipe during installation may exceed the operational stresses.
- Chemical and mechanical properties, including pipe stiffness and strength.
- Permeation resistance (i.e., in the case of potable water lines installed in contaminated soil or in soil with potential to become contaminated). Pipes laid in soils known to have gross levels of contamination are of particular concern. The American Water Works Association (ANSI/AWWA C900-97) states:

> The selection of materials is critical for water service and distribution piping in locations where the pipe will be exposed to significant concentrations of pollutants comprised of low molecular weight petroleum products or organic solvents or their vapors. Research has documented that pipe materials, such as polyethylene (HDPE), polybutylene, polyvinyl chloride (PVC), and asbestos cement (ACP), and elastomers, such as used in jointing gaskets and packing glands, may be subject to permeation by lower molecular weight organic solvents or petroleum products. If a water pipe must pass through a contaminated area or an area subject to contamination, consult with pipe manufacturers regarding permeation of pipe walls, jointing materials, etc. before selecting materials for use in that area.

- Construction and soil conditions, and related capabilities and diligence of available labor personnel for implementing bursting methods, including inspection.
- Location and external pipe environment (e.g., inland, offshore, inplant, corrosiveness of soil).
- Effluent corrosiveness, special resins, special linings, or inert materials.
- Type of burial or support (e.g., underground, exposed aboveground or elevated, underwater).
- Thermal expansion and contraction of pipe material.
- Life expectancy. The design should anticipate at least 100 years of service life (versus 50 years).
- Ease of handling and installation, including impact strength.
- Joint configuration (e.g., fused, rubber gasket, bell-less, low profile), flexibility, strength (restrained joints), and leak-tightness.
- Elongation potential during installation and postinstall relaxation time.
- In-service loads and ability of the pipe to resist such loads (e.g., collapse strength).
- Temperature or flow limitations.
- Ability to locate the pipe and vulnerability of the pipe to future unrelated construction around the pipeline.

- Tapping and other future connections to the pipeline.
- Means and practice with regard to thawing frozen service connections, where applicable.
- Ease of repair and maintenance in case of damage, including availability of repair materials and expertise.
- Practical shipping, handling, and inspection conditions associated with the jobsite.
- Adherence to governmental and other regulations and rules regarding pipe materials, including use of recycled material.

For any given project, it is extremely important that the replacement pipe material meets or exceeds the parameters defining the internal operating conditions as well as the external (in situ) conditions and is sufficiently durable to accommodate pipe bursting loads and conditions. Defining these items is critical to ensure that the new pipe system will provide adequate long-term performance for the end-user. The parameters and conditions may differ for various applications, tending to limit or define the appropriate pipe material for each case of interest. The capabilities and limitations of the different types of pipe materials available for installation by the pipe bursting method are described in the following sections.

4.1 CONCRETE PIPES

As indicated in Section 3.3.1.2, there is a wide variety of concrete pipes, including the three most commonly used: nonreinforced, reinforced, and polymer concrete. Each of these varieties may be installed by a pipe bursting/replacement operation, using appropriate procedures.

4.1.1 Nonreinforced Concrete Pipe (CP) and Reinforced Concrete Pipe (RCP)

CP, available in sizes 4 in. through 36 in., in 8-ft lengths, is intended to be used for gravity service only. RCP is available in sizes from 4 in. to more than 200 in. and has a broader range of application. Both CP (Fig. 4-1) and RCP are suitable for domestic sewers. For applications where pH values indicate a strong acidic environment that may cause degradation of the concrete surface, a corrosion protection liner should be used. Such liners may be typically installed in concrete pipe with a diameter greater than 36 in.

The design and manufacture of both CP and RCP used in trenchless applications, including pipe bursting, is subject to the requirements of *ASTM C76*.

FIGURE 4-1. Nonreinforced concrete pipe (CP).

Joints for concrete pipe are generally molded into the concrete wall and are sealed with either a rubber O-ring gasket or a bitumastic joint filler. When steel end-rings are required to reduce leakage, the steel bell and spigot shapes are secured to the pipe by welding the joint rings to the steel reinforcement. An alternative sealing method for concrete pipe uses an external steel joint collar.

Relative to other pipe materials, concrete pipe is much heavier and has inherently excellent compressive strength. These characteristics, in combination with the use of push-on joints, limit the appropriate techniques to those including static replacement methods that can apply a pushing (jacking) force to the pipe sections, maintaining compression on the assembled joints. Table 4-1 presents advantages and limitations of pipe bursting using concrete pipe.

4.1.2 Polymer Concrete Pipe (PCP)

PCP (Fig. 4-2), available in sizes ranging from 8 in. to more than 100 in. and only used for gravity service, is similar to standard CP. However, PCP substitutes corrosion-resistant polyester resin for the Portland cement component of CP, allowing application for sewers with environments within a widely varying pH range of 1 to 10. In addition, the resin

TABLE 4-1. Advantages and Limitations of Pipe Bursting Using Concrete Pipe

Advantages:	Limitations:
Long performance history.	Sensitive to nonuniform point loading in case of improper pipe jacking operations.
Thick walls provide high jacking load capability.	Poor resistance to hydrogen sulfide in sanitary sewer applications without linings.
Cartridge installation method reduces required construction site footprint.	Cracks emanating at the bell-end of the pipe can propagate and eventually undermine the structural integrity of the pipe.

binder and select aggregates provide a minimum compressive strength of 13,000 psi, which is substantially higher than that of other concrete pipe incorporating Portland cement.

Applicable standards for PCP systems include:

- *ASTM D 6783*, "Standard Specification for PCP"
- *ASTM D 4161*, "Standard Specification for Fiberglass Pipe Joints Using Flexible Elastomeric Seals"

FIGURE 4-2. Polymer concrete pipe (PCP).

- *ASTM C 579*, "Standard Test Method for Compressive Strength of Chemical-Resistant Mortars, Grouts, Monolithic Surfacing, and Polymer Concretes"
- *ASTM C 33*, "Standard Specification for Concrete Aggregates"
- *ASTM A 276*, "Standard for Stainless and Heat-Resisting Steel Bars and Shapes"

Watertight joints for PCP pipe are provided by a 316 stainless steel or fiberglass-reinforced sleeve and coupling epoxy bonded to the end of the leading pipe and utilizing elastomeric sealing gaskets on the end of the tailing pipe, consistent with the requirements of *ASTM D 4161*. The joint shall have an outside diameter equal to or slightly less than the OD of the pipe. Thus, upon assembly, the joints will be essentially flush with the OD of the pipe.

The joints for PCP are designed to transfer compressive forces through the wall of the pipe. Therefore, when used in a pipe bursting/replacement installation, the pipe must be advanced by the application of direct jacking pressure through an adaptor that will distribute the jacking pressures evenly around the joint. Systems capable of installing PCP include static systems that use a variation of direct pushing or pulling/pushing. In the direct pushing method, a jacking force is applied directly to the pipe string with the bursting head affixed to the first pipe section. Table 4-2 presents advantages and limitations of pipe bursting using PCP.

4.2 VITRIFIED CLAY PIPE

Vitrified clay pipe (VCP) (Fig. 4-3) is naturally the most chemically inert pipe material for applications in domestic or industrial gravity sewers, and is available in 4-in. through 48-in. sizes. VCP intended for

TABLE 4-2. Advantages and Limitations of Pipe Bursting Using Polymer Concrete Pipe

Advantages:	Limitations:
Excellent resistance to internal and external corrosion.	Limited performance history in the United States.
Excellent resistance to abrasion.	Sensitive to nonuniform point loading in case of improper pipe jacking operations.
Strong material in compression; high compressive (jacking) load capability.	
Cartridge installation method reduces required construction site footprint.	
Superior internal flow characteristics relative to other cement-based pipes.	

FIGURE 4-3. Vitrified clay pipe (VCP).

jacking typically has thicker walls than VCP pipes used for open cut construction, and has joints that are designed to be watertight and leak-free. The product's resistance to internal and external corrosion and abrasion, as well as its high jacking load capacity, has allowed it to become widely accepted among owners and engineers for trenchless applications, including pipe bursting/replacement.

Applicable standards for VCP include:

- *ASTM C 1208*, "Vitrified Clay Pipe and Joints for Use in Microtunneling, Sliplining, Pipe Bursting and Tunnels"
- *ASTM C 301*, "Test Methods for Vitrified Clay Pipe"
- *ASTM C 67*, "Test Methods for Sampling and Testing Brick and Structural Brick and Structural Clay Tile"
- *ASTM D 395*, "Test Methods for Rubber Property—Compression Set"
- *ASTM D 412*, "Test Methods for Vulcanized Rubber and Thermoplastic Rubbers and Elastomers—Tension"
- *ASTM D 417*, "Test Methods for Rubber Property—Effects of Liquids"
- *ASTM D 573*, "Test Methods for Rubber Deterioration—Deterioration in an Air Oven"
- *ASTM D 1149*, "Test Methods for Rubber Deterioration—Surface Ozone Cracking in a Chamber"

VCP jacking pipe uses a 316 stainless steel sleeve and coupling and elastomeric sealing gaskets on both ends of a machined spigot to provide a watertight, slip-assembled joint. When connected to adjacent pipes, the sleeve or coupling is recessed below the exterior of the pipe. Table 4-3 presents advantages and limitations of pipe bursting using VCP.

4.3 METALLIC PIPES

Metallic pipes are designed based on restricting stresses due to working and transient loads below the elastic strength of the material. The pipes of interest include DIP and steel pipe. These products have similar mechanical properties (e.g., a minimum yield strength of 42,000 psi) and are designed using similar techniques, such that the pipe remains fully elastic without degradation of mechanical properties.

4.3.1 Ductile Iron Pipe (DIP)

Ductile iron is produced by the same basic melting process that has been used for more than 150 years to make the original pit-cast gray iron pipe. The primary difference between the use of the older CI and DIP is

TABLE 4-3. Advantages and Limitations of Pipe Bursting Using Vitrified Clay Pipe

Advantages:	Limitations:
Long performance history.	Sensitive to nonuniform point loading in case of improper pipe jacking operations.
Excellent resistance to internal and external corrosion.	Short available lengths require more joints.
Excellent resistance to abrasion.	Cracks emanating at the bell-end of the pipe can propagate and eventually undermine the structural integrity of the pipe.
Strong material in compression; high compressive (jacking) load capability.	
Cartridge installation method reduces required construction site footprint.	
Low thermal expansion coefficient; does not require provision for length changes.	
Sleeve and pipe retain corrosion resistance when gouged or abraded.	
Higher temperature flows are acceptable.	

the design basis for the pipe. Because CIP has relatively low elongation capability, it is designed as a rigid pipe (taking into consideration bending resulting from external loads). In the late 1940s, it was discovered that the flake-shaped graphite in CI, when inoculated with magnesium, reformed into spheroidal or nodular graphite. This produces a much stronger iron with excellent ductility and toughness. Due to the added ductility, DIP may be designed as a flexible conduit, taking into consideration the support given to the pipe by the soil sidefill material.

DIP, in sizes 4 in. through 64 in. for either gravity service or pressure service, is designed and manufactured in accordance with *AWWA C150* and *AWWA C151*. For gravity sewer applications, the design and manufacture are also in accordance with *ASTM A 746*.

DIP has four joint configurations that are compatible with pipe bursting/replacement. Two of the joints are bell-less: a gravity service, containing a simple tongue and groove O-ring gasketed joint; and a pressure service version, acceptable to 250 psi, with an independent internal coupling that seals with twin O-ring gaskets on the two opposing bells. Both joints are machined from special thickness (class) pipe and are available in sizes 4 in. through 16 in. A third type is a conventional belled joint available in sizes 4 in. through 64 in. and internal pressures to 350 psi. The remaining design is a flexible joint, available in sizes 4 in. through 42 in., composed of a conventional bell joint with a restraining segment recessed inside a smooth-profile bell. For this flexible restrained joint, the pipe is pulled into the existing pipe with spigots oriented forward. All four joints are available in 20-ft standard joint lengths, and optional (shorter) joint lengths are available as required to meet the site conditions.

The three unrestrained joint types discussed here require direct jacking to advance the replacement pipe through the expanded existing pipe. Such DIP lines have been successfully installed by means of static systems that use a variation of direct pushing or pulling/pushing. In the direct pushing method, a jacking force is applied to the rear of the new pipe string with the bursting head affixed to the first pipe section. The static pull/push system has two variations. The first method applies a pulling force directly to the pipe bursting head and, using sectional rods preinstalled in the new replacement pipe sections, transmits the pulling force at the bursting head to a pushing force at the back of the unrestrained joints. The other technique utilizes a bursting head that is allowed to slide over or float on the pulling rods. With the pulling rods continuous from the static pulling equipment to the back of the replacement pipe, the new pipe with the floating bursting head connected to the front of the sectional pipe string is advanced through the existing pipe by an effective pushing action.

Pipe bursting using sectional DIP employs the so-called cartridge method for installation. This technique is ideal for congested areas requiring small worksite areas or where sufficient linear area is not available for preassembly of the entire pipeline on the surface and pulling it into the existing pipe as a single continuous unit.

Planning for the installation of DIP, by either open cut or trenchless construction, must include an evaluation of the soils to determine if the in situ environment will be aggressive to the DIP, resulting in the potential for external corrosion. The industry standard for corrosion protection for DIP is loosely applied polyethylene encasement. Although this system has been very successful in open cut trench construction, the use of polyethylene encasement (or any type of external coating) in a pipe bursting installation may not be practical. Therefore, corrosion control methods should be investigated, based upon the likelihood of occurrence and the consequences, as suggested by the Design Decision Model of the Ductile Iron Pipe Research Association. For example, methods similar to those described in Section 3.3.2.1, including cathodic protection, may be deployed. Figure 4-4 illustrates a pipe bursting operation with DIP, and Table 4-4 presents advantages and limitations of pipe bursting using ductile iron pipe.

4.3.2 Steel Pipe (SP)

Steel pipe, in sizes 4 in. and larger, is designed and manufactured in accordance with AWWA's *Manual of Standard Practice M11* and *AWWA C200*. SP is vulnerable to internal and external corrosion, but there are

FIGURE 4-4. Pipe bursting with DIP.

TABLE 4-4. Advantages and Limitations of Pipe Bursting Using Ductile Iron Pipe

Advantages:	Limitations:
Long performance history.	In aggressive soil environments, unprotected pipe is susceptible to external corrosion.
Widely accepted by many water utilities and installation contractors.	Cement mortar linings for DIP have limitations in aggressive sewer environments, thus requiring the application of an inert lining material, such as a ceramic epoxy.
Enhanced hydraulic capabilities; actual ID is greater than nominal size.	
Strong material, able to withstand high operation loads including: internal pressure, external loads, bearing loads, impact, and beam bending loads.	
Available for pressure, gravity, vacuum, or external pressure applications.	
Wide range of available sizes and pressure classes.	
Pipe wall is impermeable to chemicals in contaminated soil areas; special chemical- and permeation-resistant rubber gaskets are available.	
Cartridge installation method reduces required construction site footprint.	

various procedures for providing protection. These include linings and coatings such as cement mortar, paint, polyurethane, tape, coal tar enamel, and epoxy. In addition, cathodic protection may be used as an auxiliary procedure (Najafi 2005). In practice, steel has been rarely employed in pipe bursting/replacement applications.

4.4 PLASTIC PIPES

4.4.1 High-Density Polyethylene (HDPE) Pipe

PE is a thermoplastic material made by the polymerization of ethylene gas. Thermoplastics can be heated, melted, and resolidified by cooling. This permits the manufacture of pipe by continuous extrusion through a die, and pipe is manufactured in sizes up to 63 in. Most PE pipe installed by pipe bursting is of the high-density variety (HDPE).

HDPE pipe is available with iron pipe size (IPS) or ductile iron pipe size (DIPS) ODs. The pipes are classified on the basis of their dimension ratio (DR), which equals the OD divided by the minimum wall thickness. DR19 and DR17 are commonly used in pipe bursting/replacement of gravity flow lines. Pipe used for applications with high internal pressure may require lower DR values. In such cases, the required wall thickness may be greater than that of other types of pipe. However, this does not necessarily correspond to reduced flow characteristics for the same nominal size, because HDPE pipe has a relatively smooth inner surface.

The applicable standards for PE pressure service lines are *AWWA C901* and either *ASTM D 2239* (copper tubing size), *ASTM D 3035* (IPS size), or *ASTM D 2737* (ID controlled). Pipe for mains and transmission lines are generally specified in accordance with *AWWA C906* and *ASTM F 714*. *AWWA C901* and *AWWA C906* include appropriate PE resin material requirements for pipe grade materials. *ASTM F 714* does not include such a requirement but the material is specified by designating a cell class number in accordance with *ASTM D 3350*, "PE Plastics Pipe and Fittings Materials." Thus, high-quality black HDPE material for municipal water and sewer pipe has a cell class of 345464C and a grade designation of PE3408 per *ASTM D 3350*. Gray material is designated by 345464E. New materials are presently under development which will have higher cell class values.

HDPE pipe resists external pressure by the combination of its own inherent stiffness and the support the pipe receives from the surrounding soil. In addition, for pipe installed during bursting, the load on the pipe is similar to that occurring under tunnel loading, for which significant soil arching may develop in dense sands and clays. The corresponding earth load may be estimated using Terzaghi's arching coefficients, which indicate that the effective external soil pressure on the pipe is considerably lower than that corresponding to the soil immediately above, especially as the ratio of depth to pipe diameter increases (Petroff 1990).

4.4.1.1 Pipe Bursting with HDPE Pipe. British Gas developed pipe bursting in the early 1980s, splitting low-pressure CI gas lines and inserting high-pressure HDPE pipe. Shortly thereafter, the process was used to burst municipal water, gravity sewer, and force main sewer lines, which were also replaced with HDPE pipe. British Gas has reportedly installed approximately 9,000 miles of HDPE pipe by means of the bursting technique.

HDPE pipe may be installed by any of the bursting systems, including pneumatic, static pull, hydraulic expansion, and pipe splitting. Equipment is available for installation of pipe up to 54 in. in diameter. HDPE pipes through 36-in. diameter are installed on a regular basis. To facilitate inspection of gravity flow pipes, gray-colored HDPE pipe is available. Black HDPE with a white inner lining is also available at an added cost.

The insertion procedure involves fusing (Section 4.4.1.2) the HDPE pipe into the length required for insertion, stringing it on the ground, or placing it on rollers at grade level. The pipe is mechanically attached directly to the bursting tool. A small trench is usually sufficient to bring the HDPE pipe from surface grade elevation to existing pipe grade, utilizing its inherent flexibility. HDPE pipe can be cold bent to a radius of curvature of approximately 25 to 30 times the OD (e.g., DR17 has a cold radius of curvature of 27 times its diameter). Because of the relatively small radius, the preparation and assembly area for the pipe prior to insertion does not have to be aligned with the original pipe being replaced. Figure 4-5 illustrates a pipe bursting operation with HDPE.

During installation, the pull force should not exceed the allowable tensile load of the HDPE pipe. Allowable tensile loads are provided in *ASTM F 1804* and can also be obtained from the pipe manufacturer. The allowable stress is dependent upon load duration and temperature.

When the lead end of the pipe reaches the exit pit, it should be inspected for surface damage. Surface scratches or defects should be less than 10% of the wall thickness required for the service pressure. Elevated temperature during installation results in a reduction in the modulus of elasticity and may cause temporary changes in the pipe length (i.e., approximately 1 in. increase or decrease per 100 ft of pipe for each 10 °F temperature increase or decrease). Because the corresponding modulus of thermal expansion for HDPE is only one-thousandth that of steel pipe, the result-

FIGURE 4-5. Pipe bursting with HDPE pipe.

ing restraining forces are generally readily withstood. Nonetheless, it is prudent to allow the pipe to relax for 12 to 24 hours before final tie-ins and connections are made. This allows the pipe to recover (contract) from any longitudinal increase that may have occurred during pull-in (due to thermal expansion or viscoelastic effects) and will reduce the risk of pipe subsequently pulling out of the termination manholes.

A common misconception regarding the bursting method is that the existing pipe fragments from CIP or VCP could cut or damage HDPE plastic pipe during the pull-in operation. British Gas conducted a 10-year blind study of bursting CIP in which HDPE was installed using a thin-wall protective sleeve for 50% of the test sections and without the protective sleeve in the balance. No failures were observed in either category, confirming that no auxiliary mechanical protection is necessary. This is apparently due to a yielding soil trench and the tendency for CIP and VCP to fracture as slabs that are not likely to damage the replacement pipe. However, when upsizing, fragments from burst DIP may damage HDPE pipe. This is most likely to occur when the bursting is an upsize of more than one size; size-on-size bursting is not considered to be a problem. When upsizing, consult a pipe bursting expert to determine if HDPE pipe can be suitably inserted into the DIP.

4.4.1.2 Joints and Transitions. HDPE pipe and fittings may be joined by heat-fusion welding, resulting in a joint with essentially the same strength as the original pipe. During fusion welding, mating surfaces are prepared, heated until molten, joined together, and cooled under pressure, resulting in a seamless, monolithic piping system. All fusion procedures require appropriate equipment for surface preparation and alignment, as well as temperature-controlled heating irons with properly shaped, nonstick heater faces. Fusion is accomplished within pressure and temperature limits established by the PPI and published by the pipe manufacturer. Because the welded joints display the same tensile strength as the pipe, thrust blocks are not required. Most PE pipe fusion joining is done above the ditch and the pipe is then guided into the ditch after creating the proper length. However, for pipe up to 36-in. diameter, there are fusion machines capable of being inserted in a ditch to complete the final tie-in or connection to existing systems.

Electrofusion sleeves represent another option for joining HDPE pipe. Heater wires embedded in the sleeve melt the joining surfaces. Electrofusion connections are convenient for applying in trenches or in tight spaces otherwise inaccessible to fusion machines.

Sidewall fusion is a method of fusing a branch saddle or tapping tee onto an HDPE main. This process may be used to hot-tap lines and tie-in services. Electrofusion saddles are also available; some have a built-in brass transition coupling.

Mechanical couplings and transitions may be used for specific applications. A flange adapter is available for mating with a 150-lb drilling flange, and a mechanical joint adapter is available for connecting to standard DIP bells meeting *AWWA C111*. Both of these connections effectively seal against leakage and provide restraint against pull-out. Mechanical couplings, tapping saddles, and repair clamps are also available for completing installations or for maintenance purposes. Appropriate products are designated by the manufacturers, including application information such as the possible need to insert a metal sleeve or stiffener within the HDPE pipe.

When transitioning from HDPE pipe to unrestrained gasket-jointed pipe such as PVC or DIP, thrust restraint may be required. Upon pressurization, HDPE pipe contracts in length and can pull apart PVC or DIP joints located upstream or downstream from the transition connection. A thrust anchor at the point of transition or the use of mechanical restraints on the PVC or DIP pipe will prevent such incidents.

4.4.1.3 Summary. HDPE is the most commonly used material for pipe replacement. Table 4-5 presents advantages and limitations of pipe bursting using HDPE pipe.

A bursting application with HDPE replacement pipe should include the following procedures during the preliminary (design), execution (during), and conclusion (inspection) phases:

- Preliminary (design):
  - Specify replacement pipe per appropriate ASTM or AWWA standard.
  - Select size based upon actual flow capacity (function of ID and the Hazen-Williams coefficient or Manning number).
  - Select wall thickness (DR) based upon anticipated external loads, pull loads, and experience.
  - Provide the installer with the pipe's allowable tensile load.
  - Specify fusion procedures.
- Execution (during):
  - Establish a preparation and assembly area for the pipe.
  - Inspect fused connections.
  - Limit the radius of curvature of the pipe to the recommended limit.
  - Install pipe (bursting/replacement).
  - Monitor the operation.
- Completion (inspection):
  - Inspect lead end of pipe for scratches or gouges.
  - Allow time for the pipe to relax before making tie-ins.
  - Monitor the leak testing of the pipe.

TABLE 4-5. Advantages and Limitations of Pipe Bursting Using HDPE Pipe

Advantages:	Limitations:
Extensive performance history (British Gas installed 9,000 miles by bursting).	Maximum operating temperature of 140 °F.
Fused joint creates a jointless, welded pipe.	Lower hydrostatic design basis (HDB) than other thermoplastic materials, requiring greater wall thickness for specified pressure.
Resistant to a wide range of chemicals, including many highly corrosive fluids.	Special equipment and trained operators required for fusion welding.
Immune to electrochemical-based corrosion that afflicts metals.	Permeable to hydrocarbon compounds.
Invulnerable to biological attack.	Difficult to locate when buried; may require adjacent metallic tape or trace wire.
Smooth interior surface enhances flow and has a low tendency to fouling (Hazen-Williams coefficient $C = 150$).	
Freezing water will not crack pipe.	
Resistant to abrasion; suitable for use in many slurry applications.	
High flexibility; short bending radius allows for small entry pits.	
High impact resistance.	
Low surge pressures and high fatigue endurance.	

- Specify subsequent connections for transitioning to other types of pipes or structures.
- Specify HDPE or mechanical fittings for completing lateral connections.

4.4.2 Polyvinyl Chloride Pipe (PVC)

PVC is a viscoelastic, thermoplastic pipe manufactured from resin that is produced from very basic components—natural gas or petroleum, salt-water, and air. The ethylene from the natural gas or petroleum combines with chlorine from the salt to form a vinyl chloride gas that is polymerized to make PVC. The resin itself cannot successfully be extruded into rigid pipe until it is mixed with other ingredients, or compounded. These other ingredients include lubricants, heat stabilizers, ultraviolet (UV) inhibitors, processing aids, colorants, and fillers.

PVC pipe is extruded in nominal diameters of 0.5 in. to 48 in. and various dimension ratios (DRs). The most popular nominal sizes are those that have ODs that match DIP sizes. PVC pipe is also made in specific ODs for sewer applications, iron pipe sizes, and schedule pipe.

PVC is inherently resistant to attack from most chemicals; however, all thermoplastics with exposure to high concentrations of aromatic hydrocarbon compounds should be reviewed with the pipe supplier, the manufacturer, or both regarding possible chemical permeation.

PVC has a tensile strength of 7,000 psi and an HDB of 4,000 psi for pressurized applications. This results in a stronger pipe than HDPE, allowing thinner walls and a smaller OD, a larger internal flow area, or both. For pipe bursting applications, fusible-joint PVC pipe is now available (Section 4.4.2.1). This flush, gasketless joint provides a high ratio of flow area to overall outer pipe dimension, resulting in a smaller expansion necessary to accommodate the new joined pipe. The fused joint also displays a tensile capability equal to that of the original PVC pipe, facilitating pull-in of the replacement pipe.

The thermal coefficient of expansion for PVC is 3.0×10^{-5} in./in./°F. This equates to a 0.3 in. increase or decrease per 100 ft of pipe for each 10 °F temperature increase or decrease (less than one-third that of HDPE) and is therefore also not a significant issue during handling or installation. In particular, when installing PVC in any trenchless application, the time required for the pull-back is generally sufficient for the pipe to reach a thermal equilibrium with the surrounding soils.

Immediately before lowering the pipe into the trench, each joint of PVC pipe shall have the entire exterior pipe surfaces checked for scratches, gouges, or defects. Any section found to have surface imperfections greater than 10% of the wall thickness shall be removed from the project site. In addition, when the lead end of the pipe reaches the exit pit following installation, it should similarly be inspected for surface damage. Surface scratches or defects should be less than 10% of the wall thickness.

The design and manufacture of PVC pipe for water service is governed by *AWWA C900, AWWA C905,* and *AWWA C605* as well as others, as follows:

- *AWWA C900,* "AWWA Standard for Polyvinyl Chloride (PVC) Pressure Pipe and Fabricated Fittings, 4 in. through 12 in. (100 mm through 300 mm), For Water Distribution"
- *AWWA C905,* "AWWA Standard for Polyvinyl Chloride (PVC) Pressure Pipe and Fabricated Fittings, 14 in. through 48 in. (350 mm through 1,200 mm), For Water Distribution"
- *AWWA C605,* "AWWA Standard for Polyvinyl Chloride (PVC) Underground Installation of Polyvinyl Chloride (PVC) Pressure Pipe and Fittings for Water"

- *ASTM D 1784*, "Rigid Poly (Vinyl Chloride) (PVC) Compounds and Chlorinated Poly (Vinyl Chloride) (CPVC) Compounds"
- *ASTM D 1785*, "Poly (Vinyl Chloride) (PVC) Plastic Pipe, Schedules 40, 80, and 120"
- *ASTM D 2241*, "Poly (Vinyl Chloride) (PVC) Pressure Rated Pipe (SDR PR Series)"
- *ASTM D 3915*, "Rigid Poly (Vinyl Chloride) (PVC) and Chlorinated Poly (Vinyl Chloride) (CPVC) Compounds for Plastic Pipe and Fittings used in Pressure Applications"
- *ASTM F 679*, "Poly (Vinyl Chloride) (PVC) Large Diameter Plastic Gravity Sewer Pipe and Fittings"
- *PPI TR-2, Plastic Pipe Institute Technical Report 2*, "PPI PVC Range Composition of Qualified Ingredients"

AWWA C900 and *AWWA C905* refer to DIP sizes. Figure 4-6 illustrates pipe bursting with PVC pipe.

4.4.2.1 Joints and Transitions. PVC pipe is available in several configurations, including conventional bell-and-spigot rubber-gasketed joint,

FIGURE 4-6. Pipe bursting with PVC pipe.

coupling and spigot with restraining spline, and fusible PVC pipe. In addition, a fourth type—integrally restrained bell-and-spigot gasketed joint—has recently been developed.

Conventional bell-and-spigot PVC pipe allows for horizontal alignment changes, vertical alignment changes, or both by opening and closing of the rubber-gasketed joint, requiring little or no longitudinal bending of the pipe. Fusible PVC joints do not deflect; therefore, horizontal or vertical alignment changes require that the pipe be subject to bending. The amount of bending recommended for fusible joint PVC corresponds to a minimum radius of curvature of 250 times the actual OD of the pipe.

Assuming that two conditions are met, two lengths of PVC pipe can be fused together by the application of heat, allowing the full mechanical properties of PVC to be utilized in trenchless installations. First, the PVC pipe must be of the proper formulation to allow fusion to produce joints as strong as the pipe. Second, the specified procedure for the fusion process must be utilized.

The butt-fusion process uses industry-standard equipment, with appropriate modifications for application to PVC. The preparation, heating, and fusion use different temperatures and pressures than those required for other thermoplastics.

Fusible PVC pipe used in trenchless applications can be joined to the existing piping systems using mechanical joints with standard PVC restraints. Unlike HDPE, which is more flexible, PVC does not require stainless steel stiffening inserts for mechanical joint or other compression-type connections.

Service connections are added to a line by using a standard PVC tap saddle and tap procedure. Direct tapping is not recommended. Electrofusion sleeves or saddles are not needed for such applications.

4.4.2.2 Pipe Loading. The development of restrained as well as fusible PVC joints has necessitated the evaluation of tensile force and bending capability of appropriately joined PVC pipe. In general, the manufacturers or suppliers can provide relevant information regarding the tensile and bending characteristics of joined PVC pipe. In particular, the safe pulling force (SPF) has been measured after extensive testing of tensile capabilities of fusion joined pipe. The PVC fusion joint was determined to be as strong as the pipe itself. The SPF was determined at 73 °F, as per ASTM D 638 for temperature requirement, so installation at possible elevated temperature should be considered. The mechanical properties of all thermoplastics are dependent upon temperature; the strength decreases as temperature increases.

PVC provides a high load capability for situations in which external loads may be important, such as nonpressurized applications (including

TABLE 4-6. Advantages and Limitations of Pipe Bursting Using PVC Pipe

Advantages:	Limitations:
High HDB value allows relatively thin pipe wall thickness, maximizing available internal flow area.	Less flexible than other thermoplastics, resulting in larger bend radii.
Resistant to a wide range of chemicals, including many highly corrosive fluids.	Requires special equipment and material formulated for butt-fusion.
Butt-fused joints create a continuous, gasketless pipe.	Difficult to locate when buried; may require adjacent metallic tape or wire.
Fully restrained joints; require no thrust blocks.	Pipe must be inspected for gouges after burst installation.
Resistant to abrasion (internal and external).	
Low internal friction enhances flow characteristics.	
Low expansion and contraction.	
Readily available saddles and fittings for taps and repairs.	
Light weight in smaller diameters.	

gravity flow). For gravity flow applications, PVC stiffness is also very helpful because it enables the pipe to avoid local sags or pools in which biological growth could form, leading to potential problems for sewer applications. Table 4-6 presents advantages and limitations of pipe bursting using PVC pipe.

PART 5

DESIGN AND PRECONSTRUCTION PHASE

5.1 FEASIBILITY AND RISK ASSESSMENT

All construction projects include some elements of risk. Most underground and pipeline construction projects generally have risks associated with unknown subsurface conditions. Because pipe bursting is a rapidly evolving technology, additional risks due primarily to the relative lack of experience in similar conditions should also be considered.

The application of pipe bursting is usually advantageous over direct excavation methods only where site constraints exist. Some of the constraints that specifically justify the use of pipe bursting include:

- Deep trench excavations
- Unstable soils, groundwater
- Crowded utility corridors
- High traffic control or limited construction interruption requirements
- Significant pavement or surface restoration needs
- Hazardous soils
- High-public-impact or high-visibility locations.

Under these circumstances, the combination of unknowns related to site conditions and the applicability of construction methods and materials needs to be carefully reviewed to confirm the feasibility of using pipe bursting techniques. The planning phase activities presented in Part 2 included a range of items and investigations needed to properly evaluate a pipe bursting project, whether for gravity or pressure pipe applications. In evaluating the feasibility of successfully constructing a pipeline using pipe bursting techniques, a review of the planning phase activities should be undertaken to assess how well-defined the project is and what the potential risks to the project are. From a risk assessment perspective, the review should focus on the following considerations:

- How well are existing site conditions understood?
  - Are as-built surveys available and complete? Are existing pipe materials, fittings, repair sleeves, and concrete encasement accurately documented? Is the condition of the existing pipe compatible with improvement plans?
  - Have all existing utilities been identified and located?
  - Are soil, water, and related ground conditions understood? Do ground conditions vary significantly? Are subsurface soils or water hazardous?
- What are the construction requirements?
  - Is there demonstrated experience of successful applications with the pipe bursting technique and material planned under similar conditions?
  - Are the installed accuracy limits realistic and achievable?
  - Have all permitting requirements been identified? Who will be responsible for obtaining permits? Is permit approval a "sure thing"?
  - Have salvage and disposal needs been defined?
  - Are restoration needs clear? What measures to document existing conditions will be implemented? What will be done to monitor and protect against damage?
  - Are certified, qualified, and experienced contractors available and willing to bid for project?
  - Is there a "Plan B"? Should the work include provisions for equipment rescues and direct excavation alternatives?

Successful projects identify and define risks and anticipate measures to address and mitigate them to minimize their impacts. Because of the dynamic nature of the technology, pipe bursting projects require a strong partnership between owners, engineers, contractors, and equipment and pipe material suppliers to be successful. The need to address the project as a team is crucial to addressing the challenges that will occur as the work progresses. Identifying and developing a realistic plan to manage and share risk appropriately is an important part of effectively communicating responsibilities, defining roles, and building a strong team.

5.2 DESIGN PARAMETERS

As previously noted, the continual evolution of the pipe bursting industry makes the need for close coordination between owners, designers, equipment and pipe material suppliers, and contractors especially important. Understanding the changes and impacts of equipment improvements and pipe material advances should not be overlooked and it can be expected that pipe bursting technologies will continue to progress.

In terms of hydraulic parameters and materials of construction, design considerations for a pipe bursting project follow conventional design practice and should address the following:

- Establish design operating flows, pressures, and temperatures.
- Document how much flows will vary (minimum/peak, startup/ultimate, seasonal).
- Select materials of construction requirements (internal and external corrosion protection, abrasion and wear considerations, design and allowable thrust, and pipe loading parameters). Refer to the discussion of materials in Part 3.

Other significant design parameters, such as launch and receiving pit locations, layouts, sizes, and maximum pipe burst lengths, are functions related primarily to the pipe materials and equipment systems capabilities as well as to the project site conditions. These parameters need to be closely coordinated with contractors and confirmed during the bid submittal process. In addition, the adequacy, experience, and capabilities of potential pipe bursting equipment suppliers for constructability need to be considered.

After establishing these basic parameters, the development of a design package for a pipe bursting project becomes a specialized application that may be unique or may be compatible with a wide variety of materials and methods, depending on site constraints and owner preferences. The development of contract documents further defines the requirements of the work based on these constraints and preferences.

5.3 CONTRACT DOCUMENTS

5.3.1 Scope of Work and Special Conditions

The contract documents provide information about the scope of work and special conditions of the project. Although a conventional design-bid-build approach can be used to deliver a pipe bursting project, the proprietary nature of the materials and methods used, the contractor's experience, and the continually evolving nature of pipe bursting technology requires that the preparation of contract documents be tailored to site conditions. Similarly, because of the peculiarities associated with the pipe bursting process, alternative project delivery methodologies (e.g., design-build) have gained increasing use.

This is not to say that the contract documents for a pipe bursting job will necessarily be more complex than for a conventional pipeline design project. In many cases, this means defining the limits of the workspace from a different perspective; this generally involves shifting the importance of and emphasis on technical submittals during the bid and construction process. It may also mean that the responsibility for confirming various design details may shift from the engineer to the contractor. The following narrative has been prepared to identify the general requirements for drawings and specifications. The responsibility for and timing of preparing and submitting these items will depend on the project delivery method utilized. It should also be noted that, due to the nature of underground construction projects, there are many risks involved and therefore a partnering approach between the contractor and the owner to share the risks can lower the actual costs of the overall project.

5.3.2 Drawings

Drawings should provide information about the existing site conditions and the construction to be accomplished, including:

- Limits of work; horizontal and vertical control references
- Topography, planemetrics, and survey points of existing structures
- Boundaries, easements, and rights-of-way
- Existing utilities, sizes, locations, and materials of construction
- Plan and profile of the design alignment
- Location and size of launching and receiving pits
- Material and equipment layout and storage areas
- Launching and receiving pit details
- Details for connections to the existing piping system
- Restoration plans.

Drawings may also include information to show erosion and sediment control requirements, flow bypassing plans, service connection and reinstatement details, and support permitting activities.

5.3.3 Technical Specifications and Submittals

Technical specifications supplement the drawings in communicating project requirements. As noted earlier, technical specifications may take the form of outlining requirements that are confirmed or further detailed by submittals transmitted during the bidding and construction process.

Information to be included in the technical specifications or required in submittals should include:

- General
  - Minimum requirements for experience, certification/training, and references of the contractor, key construction personnel, materials, and equipment to be used
  - Project schedule
  - Permitting matrix and responsibilities
  - Field safety plan
- Pipe materials (also refer to Part 3)
  - Standards and tolerances for materials, wall thickness and class, testing and certification requirements
  - Material manufacturer's experience history with similar work
  - Construction installation instructions for pipe joining and handling
  - Fittings, appurtenances, and connection-adaptors
  - Requirements for annular space sealing
- Construction considerations (refer also to Part 6)
  - Flow bypassing, downtime limits, and service reinstatement requirements
  - Spill and emergency response plans
  - Traffic control requirements
  - Erosion and sediment control requirements
  - Existing conditions documentation (e.g., photographs, videos, interviews)
  - Earth support systems
  - Ground movement monitoring and existing structure and utility protection plan
  - Accuracy requirements of the installed pipe
  - Daily construction monitoring reports
  - Field testing and follow-up requirements for pipe joining, pipe leakage, disinfection, backfill, annular space sealing
  - Site restoration and spoil material disposal requirements.

5.4 GEOTECHNICAL CONDITIONS

Geological and geotechnical information related to the project site should be evaluated and detailed in the geotechnical report. Information including the type of soil, groundwater conditions, water table, proximity, and potential impacts to the existing utilities and structures should be reviewed (Section 2.1.3). The technical design should include considerations for reaction or thrust loads imposed on the working shaft walls (when applicable) so that adjacent structures and utilities are not damaged during pipe bursting operations. Settlement and heave of adjacent utilities and critical features should be evaluated.

5.5 DIFFERING SITE CONDITIONS

If site conditions are significantly different from that described in the contract documents and the contractor can show that the different conditions impact the work, the contract value should be adjusted. Preliminary investigations, such as utility mapping and geotechnical surveys, should reduce the risk many of the common problems occurring during the pipe bursting. A risk-sharing strategy, such as one recommended by the ASCE "Geotechnical Baseline Report," (ASCE, 1997) is strongly recommended. Soil conditions (i.e., rock, clay, sand, water table), when reasonably described, indicated, or implied in the contract documents, are usually not grounds for a Change of Conditions claim when using pipe bursting. Valid claims for differing underground conditions should generally be limited to unknown items (e.g., unexpected changes in trench geometry, undocumented encasement and repair or sleeves, structures, or cavities). The contract documents should contain a clause requiring written notification of claim for Change of Conditions, within certain number of days from the date of first discovery, after which no claim will be allowed.

5.6 DISPUTE RESOLUTION

Because there is always some probability of encountering unforeseen conditions underground, it is imperative that a well-devised dispute resolution plan be included in the contract document. A Differing Site Conditions (DSC) clause should be included in the contract to allow the contractor to be compensated for extra costs involved without being forced into a breach of contract. It is in everyone's best interest to resolve conflicts quickly, fairly, and equitably.

PART 6
CONSTRUCTION PHASE

6.1 WORK PLAN

It is generally the responsibility of the utility owner to perform an inspection of the existing pipe utilizing CCTV or other acceptable method prior to bidding. The results of the inspection must be made available with sufficient time for the contractors to actually evaluate the inspection prior to bidding. The inspection should be completed utilizing standard guidelines (such as those as provided by NASSCO) to provide reproducible results. The inspection results should include nomenclature of defects with descriptions (a glossary). The inspection should include name(s) of defect(s), quantifiable measurement of defect(s), location of each defect, and when the inspection was completed. The data and stationing should be recorded on the project drawings. Audio description of inspection data should also be provided for each section of existing pipe. The finished inspection data should provide a complete record along the length of pipe to be burst. The project owner should also provide the contractor with any as-built drawings for the existing pipeline and nearby facilities, geotechnical reports, and the results of any exploratory excavations that were completed to ensure the pipeline is appropriate for pipe bursting. The owner should note on the contract drawings any area where the contractor is not permitted to locate an excavation.

A work plan for pipe bursting operations should consist of two parts: the main plan and a contingency plan. The following elements should be covered in the main plan:

- Permits
- Traffic control
- Excavation and shoring, as required

- Protection of adjacent structures and utilities
- Dewatering design (if necessary)
- Bypassing (if necessary)
- Supervision and inspection
- Public safety
- Spoil removal
- Scheduling, including work days and work hours.

The contractor should be responsible for the preparation of contingency plans. The contingency plan should include symptoms of problems, such as location of equipment or materials, bypassing requirements, pit excavation, permit impacts, and schedule impacts. The contractor should submit the contingency plans required in Section 6.8 for review and acceptance by the owner before commencing work.

If there is a point of rejection during the operation, the rejection reasons should be identified and corrective actions should be planned in advance to resolve the situation without extensive delays. If an obstruction can be removed by the open cut method, then pipe bursting can continue with the same equipment. Obstructions may increase ground vibrations, which can cause ground movements and heaves. Contractors should anticipate and properly mitigate possibilities of obstructions and ground movements if adequately characterized in the contract documents.

The contractor, the owner, and the engineer should meet and jointly review the detailed schedule of activities and ensure that the schedule meets the contractual requirements. The contractor should revise the schedule of activities to address any necessary adjustments regarding the value allocation or level of detail.

In scheduling pipe bursting operations, working hours should not be unnecessarily restricted. Multiple or extended shifts should be permitted where possible. This decision should take into consideration noise levels that should not exceed the maximum levels contained in local ordinances.

6.2 WORKSPACE

Workspace is composed of the pit, the area around the pits, and the pipe laydown and buildup area, if required. These areas should provide enough room for safe operation of the equipment. All areas should be well organized to provide for safe and productive work.

Space inside of the pits should be adequate to provide for easy lifting and safe pulling or jacking operations. The entry or insertion pit should be long enough to allow the bursting head to be aligned with the existing pipe and the new pipe string to bend without any negative impact on the pipe. The manufacturer of the new pipe should be consulted to determine the minimum safe radius of curvature of the new pipe. Sharp or jagged edges at entry locations, along with any condition that may damage the new pipe, should be removed or protected so as to prevent damage to the new pipe. The pipe laydown and buildup area extends from the back of the launch pit for the length of the pipe to be burst if required by the contractor's method.

Engineers as well as contractors should consider that some new pipe material could be handled more easily than the others, providing solutions to on-site workspace restrictions (also refer to Part 4). This fact should be considered in the proposals on the new pipe material selection.

6.3 JOBSITE LAYOUT

The jobsite layout should be organized according to the plan submitted by the contractor and approved by the owner, the engineer, or both. Plan and profile of pipe bursting operation, as well as traffic control and existing utilities should be shown on the layout. The adequate performance of this plan is the responsibility of the contractor and is to be enforced by inspectors. The pipe bursting equipment, bypass system, lubrication equipment, and the new pipe sections should be located close to the pit(s) where they will be needed. Before finalizing the plan for job layout and start of the work, the local authority having jurisdiction over the project location should be notified to obtain rights-of-way and easement permits.

6.4 INSERTION (ENTRY OR LAUNCHING) AND RECEIVING (PULLING) PITS

Insertion (new pipe entry point) and receiving pits should be strategically and safely located to reduce the overall excavations on a project, considering the traffic flow and specific conditions of the project. Normally, water main systems have clusters of gate valves at street intersections and fire hydrants at intervals of 500 ft or less. Trunk sewer systems commonly have manholes at intervals of 400 to 500 ft or less. These should be prime locations for pits because valves, hydrants, and manholes are usually replaced with the pipe.

Figure 6-1 illustrates a typical pit layout for pipe bursting using general design ratios. In general, the material, diameter, and diameter thickness ratio (DR) of the new pipe will determine the length of the entry pit required. For example, the required pit length for DR17 HDPE pipe less than 18 in. in diameter is 2.5 to 3 times the pipe invert depth. HDPE DR17 pipe of diameter 18 in. and larger requires a somewhat longer pit than indicated by this general design ratio. DI and PVC pipes are relatively

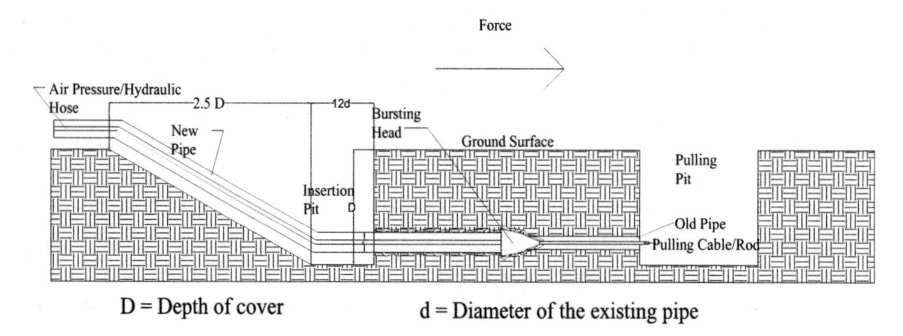

FIGURE 6-1. Pipe bursting operation layout and insertion pit.

rigid and may require significantly longer entry pits in order to minimize the imposed curvature. Overbending the pipe can cause overstressing the pipe material and create damage which may not become apparent until the pipe is placed in operation. The width of the pit is dependent upon the pipe diameter and OSHA confined space or shoring requirements. The use of appropriate pit shoring is defined by depth and ground type as contained in OSHA regulations.

Entry and receiving pits should be constructed according to the submitted drawings, which show excavation locations, dimensions, method of shoring—approved by a professional engineer—and dewatering (for both surface and groundwater), adjacent utilities, and traffic control. Utilization of the existing manhole structure as a receiving pit should be verified for the structural capacity of manhole walls to withstand the installation forces. Figure 6.2 presents an example of trench shoring.

6.5 BID SUBMITTALS

The engineer should provide the contractor with flexibility to establish the installation procedure (i.e., installation lengths, pit locations) but, rather, require a work plan prepared by the contractor to be approved by the engineer, the owner, or both. The specification should not be prescriptive in locating pits except in areas that the contractor cannot excavate or locate equipment, i.e., major intersections, hospital access, or fire station access. The contractor should submit the following information (also refer to relevant discussion in Section 5.3.3, Technical Specifications and Submittals).

- Proposed type and size of equipment, operational requirements such as air pressure and cubic feet per minute, bypass pumping or

FIGURE 6-2. An example of trench shoring.

temporary services system, dewatering system (if required), lubrication system, and pit layout and protection system.

- Description past experience with similar projects (as defined by pipe upsize percentage, existing pipe material, and new pipe material) with contact information of owners or engineers who have successfully used a similar system by the same manufacturer, the same contractor, or both.
- In case the contractor does not have sufficient experience with pipe bursting, a written commitment from the equipment manufacturer to provide an experienced technician to oversee the operation should be required.
- Description of the means and methods to locate service laterals contained within the burst, disconnect service laterals before the burst, reconnect service points following the burst, and reinstating service.
- Description of the method to remove spoils.
- Permits for disposal of any spoils.
- Description of the lubrication system (if lubrication is needed), including the material safety data sheets (MSDSs) of lubricants, type of lubricants (bentonite or polymers or both), and proportions of mixtures. Additives must have approvals by NSF International (NSF) per *NSF/ANSI Standard 60* Certified.

- An analysis of new (replacement) pipe push or pull loads with an appropriate safety factor, considering the allowable tensile or jacking strength of the new pipe. The new pipe supplier should be consulted to determine pulling or jacking loads and for determination of a safety factor. The estimate shall include pipe upsize percentage as a factor. (Pipe upsize is a measure of host pipe ID and replacement pipe OD.)

6.6 LUBRICATION

The purpose of lubrication is to reduce friction between the new pipe, the existing pipe and pipe bedding mixture, thereby reducing the pulling or pushing forces on the new replacement pipe. Lubrication should be used according to the contractor's plans to ensure successful completion of the job. The contractor should use a lubrication design recommended by the lubrication manufacturer and approved by the owner, the engineer, or both.

The soil condition should dictate the correct lubrication additives for each pipe bursting operation. Basically, bentonite is used in coarse soils (sand and gravel) whereas a mixture of bentonite and polymers may be used in the fine and clayey soils. Other additives with specific uses could be applied in pipe bursting operations, such as:

- Detergents, as wetting agents in clays
- Torque-reducing agents to assist in lubrication and stabilizing of ground during upsizing
- Filter control additives to help reduce fluid loss in the formation, and to prevent swelling.

The lubrication fluid supplier may be consulted for the fluid mix components, proportions, and mixing procedures for the specific soils or trench conditions surrounding the pipe. The contractor should be aware that soil conditions could change along the length of the existing pipe and this may require additional additives to be available and added to the lubrication in order to achieve successful lubrication.

Usually a lubrication mixer is used to mix lubricants and possibly a pump is used to distribute lubricant in the annular space surrounding the replacement pipe. The lubrication fluid is circulated to the exterior of the pipe through opening(s) directly behind the head or expander. Lubricants are recommended in the following cases:

- Increasing two incremental pipe sizes or more and burst length exceeding 300 ft
- Diameter of new pipe exceeds 12 in.
- Existing pipe is under groundwater
- Free-flowing soil conditions

- As recommended by the pipe bursting equipment manufacturer, due to specific site and project conditions
- Expansive clay soils.

6.7 DEWATERING

Dewatering may be required at the insertion and receiving pits when the pipe profile is below the groundwater table. Depending on the soil type, the dewatering system may be a sump pump, a well point, or a vacuum well system.

6.8 CONTINGENCY PLAN

The contractor should be responsible for the preparation of a contingency plan, which must be reviewed by the owner, the engineer, or both before commencing the work. The contingency plan should cover the following potential existing pipe conditions:

- Structural problems or collapse of existing pipe
- Existing pipe designation is different from actual pipe in ground
- Obstructions and unexpected interferences
- Previous point repairs with different materials (such as encased concrete)
- Loss of lubricants
- Excessive pit wall movement during the static pull
- Excessive settlement or heave
- Contaminated ground.

If there is a possibility of rejection point or obstruction during the operation, the rejection reasons should be identified and corrective actions should be planned in advance to resolve the situation without extensive delays. If an obstruction can be removed by the open cut method, then pipe bursting can continue with the same equipment. Obstructions may increase ground vibrations, which can cause ground movements and heaves. Contractors should anticipate and properly mitigate possibilities of obstructions and ground movements, if ground conditions are adequately characterized in the contract documents.

6.9 INSPECTION AND MONITORING

A qualified and trained inspector in pipe bursting operations should inspect the project to verify it has been constructed according to approved submittals, technical specifications, and contract documents. The inspector should check the new pipe upon arrival at the job site to make sure it meets manufacturer specifications such as type and wall thickness, diameter, and resulting DR number. The contractor should submit final inspection and testing results before requesting final payment. For pressure pipe applications, a pressure test should be performed to verify connections. For gravity installations, a low-pressure air test should be performed according to the procedures described in *ASTM C 828, ASTM C 924, ASTM F 1417*, or other appropriate procedures.

6.10 AS-BUILT DRAWINGS AND DOCUMENTATIONS

The engineer's design drawings showing installed pipe locations and elevations (alignments and profiles) should be verified against the actual installation. Surface elevations should be documented if the area is at risk of utility interference, ground heaving, or structural damage to pavement, buildings, and facilities in close proximity of new pipe alignment.

The following information should be documented in pipe bursting projects:

- Documentation of insertion and receiving pits
  - Information about the number of manholes, catch basins, and pits along the existing pipe alignment should be presented.
- Surface monitoring system
  - If necessary, surface survey monitoring points must be established to check possible settlement and heave. Local department of transportation (DOT) and railroad guidelines should be considered and followed.
  - The postinstallation survey should include information about new pipe (such as post CCTV) as well as existing utilities.
- Other parameters that should be documented are:
  - New pipe testing results
  - Pulling or pushing forces
  - New pipe alignment and profile as compared to the existing pipe
  - Lubrication used
  - Removal of contaminated soil or existing pipe, if applicable.

6.11 RECONNECTION OF SERVICE

In case a PE pipe is used as a new pipe, the reconnection of service, sealing of the annular space at the manhole location, or backfilling of the insertion pit for new pipe must be delayed for the manufacturer's recommended time, but normally not less than 4 hours. This period allows for PE pipe shrinkage due to cooling and pipe relaxation owing to the tensile stresses induced in the pipe during installation. Following the relaxation period, the annular space in the manhole wall may be sealed. Sealing is extended a minimum of 8 in. into a manhole wall in such a manner as to form a smooth, watertight joint. Ensuring a proper bond between the replacement pipe and the new manhole wall joint is critical.

Service connections can be reconnected with specially designed fittings by various methods. The saddles, made of a material compatible with that of the new pipe, are connected to create a leak-free joint. Different types of fused saddles (electrofusion saddles, conventional fusion saddles) are to be installed in accordance with manufacturers' recommended procedures.

Connection of new service laterals to the pipe also can be accomplished by compression-fit service connections. After testing and inspection to ensure that the new replacement pipe meets all the required specifications, the pipeline returns to service. Figure 6.3 illustrates service lateral excavations.

6.12 MEASUREMENT AND PAYMENT

Payment for pipe bursting may include separate additional items such as insertion pits and receiving pits, locating and potholing utilities, exca-

FIGURE 6-3. Service lateral excavations.

vation and reconstruction of service laterals, any point repair, obstruction removal, stormwater pollution prevention and pit erosion control planning, site restoration, existing and new pipe inspections, bypass pumping or temporary services, removal of hazardous materials, mobilization, demobilization, dewatering, and other activities. The cost of the new pipe installation includes: cost of the pipe material; cost of pipe bursting installation, including bursting existing pipe and inserting new pipe; sealing materials at manholes and annulus (if required); and posttesting. Payment items may include open trench excavation because some reaches may require open trench excavation to complete a portion of the project. The following are potential payment arrangements:

- Pipe installation by pipe bursting should be paid per linear foot and its price should depend on size of the pipe (diameter, length, type of material, quantity, and depth).
- Location, connection, and reconstruction of services should include fittings and pipe, and it may be paid for per each connection made.
- The payment of the point repairs should be established on a case-by-case basis.
- The payment for obstruction removal should be made depending on the scope of the work, and the cost should be evaluated on a case-by-case basis.
- Site restoration may be paid per square foot, per location, or by lump sum, depending on the project requirements.
- Internal inspection, cleaning, disinfection, and pressure testing may be included in the cost of the installed pipe.
- Bypass pumping or temporary service connections may be included as part of the pipe installation, whereas CCTV inspection, repair, or related work may be treated as additional work and paid for separately.
- Manhole replacement may be paid by the item or by vertical foot.

It should be noted that it is common for many of the above items to be considered as incidental to the unit of new pipe installed and thus no separate pay items would be established for them.

6.13 TYPICAL COSTS FOR PIPE BURSTING

As for any construction project, pipe bursting costs vary with project conditions, such as metropolitan location, type and size of existing and new pipe, number and location of laterals, project size, surface and subsurface conditions, and existing utilities. It is very important that the engineer prepares a complete set of bid documents with major bid items indicated and that the contractor understands conditions of the project in order to submit a responsible and responsive bid. In this way, comparable bids are received from all contractors and change orders after the bid is accepted are minimized.

6.14 SAFETY ISSUES

The contractor is responsible for safety on the jobsite. Safety should be covered according to the relevant OSHA standards. Special attention should be paid to safety of workers, pedestrians, and the traveling public during the entire pipe bursting project. Common safety issues may include trench and pit shoring, vehicular traffic, existing utilities, and overhead power lines.

PART 7

REFERENCES

ASCE (1997). "Geotechnical baseline reports for underground construction," American Society of Civil Engineers, Reston, VA, *ASCE*, 0-7844-0249-3, 1997.

ANSI/AWWA C900-97 (Revision of ANSI/AWWA C900-89), "AWWA standard for polyvinyl chloride (PVC) pressure pipe and fabricated fittings, 4 in. through 12 in. (100 mm through 300 mm), for Water Distribution," Section 4.1—Permeation, Page 3.

Boot, J., Woods, G., and Streatfield, R. (1987). "On-line replacement of sewer using vitrified clayware pipes." *Proc., No-Dig International '87*, London, UK.

Fisk, A. T., and Zlokovitz, R. (1992). "Replacement of steel gas distribution mains with plastic by bursting." *Proc., No-Dig International '92*, Washington, D.C.

Fraser, R., Howell, N., and Torielli, R. (1992). "Pipe bursting: The pipeline insertion method." *Proc., No-Dig International '92*, Washington D.C., North American Society for Trenchless Technology, Arlington, VA.

Howell, N. (1995). "The polyethylene pipe philosophy for pipeline renovation." *Proc., No-Dig International '95*, Dresden, Germany, ISTT, UK.

Najafi, M. (1994). "Trenchless pipeline rehabilitation." Trenchless Technology Center, Louisiana Tech University, Ruston, La.

Najafi, M. (1999). "Overview of pipeline renewal methods." *Proc., Trenchless Pipeline Renewal Design & Construction '99*, University of Missouri–Kansas City, Mo.

Najafi, M. (2005). *Trenchless technology: Pipeline and utility design, construction and renewal*, McGraw-Hill, New York.

North American Society for Trenchless Technology (NASTT). (2004). "Pipe bursting good practices guidelines," NASTT, Arlington, VA.

Petroff, L. J. (1990). "Review of the relationship between internal shear resistance and arching in plastic pipe installations." *Buried plastic pipe technology*, G. S. Buczala and M. J. Cassady, eds., ASTM International, West Conshohocken, Pa.

Topf, H. (1991). "XPANDIT trenchless pipe replacement." *Proc., North American No-Dig '91*, Kansas City, MO.

Topf, H. (1992). "XPANDIT trenchless pipe replacement." *Proc., No-Dig International '92*, Washington, D.C. North American Society for Trenchless Technology, Arlington, VA.

Tucker R., Yarnell, I., Bowyer, R., and Rus, D. (1987). "Hydraulic pipe bursting offers a new dimension." *Proc., No-Dig International '87*, London, UK.

INDEX

Note: Entries from figures are followed by *f*; tables are followed by *t*